guesstimation

* * * * * * * * * * * * * * * * * * *

Lawrence Weinstein and John A. Adam

* * * * * * * * * * * * * * *

guesstimation

✳ ✳ ✳ ✳ ✳ ✳ ✳ ✳ ✳ ✳ ✳ ✳ ✳ ✳ ✳ ✳

Solving the World's Problems
on the Back of a Cocktail Napkin

Princeton University Press   Princeton and Oxford

Published by Princeton University Press, 41 William Street, Princeton, New Jersey 08540

In the United Kingdom: Princeton University Press, 3 Market Place, Woodstock, Oxfordshire OX20 1SY

Library of Congress Cataloging-in-Publication Data

Weinstein, Lawrence, 1960-

Guesstimation: solving the world's problems on the back of a cocktail napkin / Lawrence Weinstein and John A. Adam.

p. cm.

Includes bibliographical references and index.

ISBN 978-0-691-12949-5 (clothbound : alk. paper) 1. Estimation theory–Miscellanea. 2. Problem solving–Miscellanea. I. Adam, John A. II. Title.

QA276.8.W45 2008

519.5′44–dc22   2007033928

British Library Cataloging-in-Publication Data is available

This book has been composed in Minion with Scala Sans display

Printed on acid-free paper. ∞

press.princeton.edu

Printed in the United States of America

10   9   8   7   6   5   4   3

To my wife, Carol, and my children, Lee and Rachel,
who understand when I get that vacant look
in my eyes and start mumbling exponents.   [LW]

To four wonderful and cherished nonmathematicians:
my wife, Susan, and my children, Rachel, Matthew, and
Lindsay, and last, but not least (though he is still very
small), to my grandson, John Mark—may you come to
love numbers at least as much as I do!   [JAA]

Contents

Acknowledgments

We would like to thank everyone who helped us, or, at least, who did not slow us down too much.

I (John) would like to thank my Department Chair, J. Mark Dorrepaal, for his enthusiastic support for this venture (and many others). Our editor, Vickie Kearn, once told me that writing a book was like giving birth to a pine cone; her sound advice and sense of humor has enabled this "baby" to be brought to full term in a relatively pain-free fashion (though with occasional cravings for ice cream). My coauthor, Larry, has been a pleasure to work with, and has kept me on my intellectual toes (without treading on them). My delight at "discovering" Patty Edwards and her artistic talents, as she walked her dog and I just walked around our neighborhood, has increased daily. Her superb contributions to this book will no doubt engender a whole new genre of stories: "A physicist, an artist, and a mathematician walked into a bar..." Finally, my wife, Susan, has been a wise and effective sounding board for my book-related mutterings, and I thank her for all her love and advice.

In particular, I (Larry) thank the Old Dominion University Physics Department, who encouraged me to develop the course "Physics on the Back of an Envelope," on which this book is based. Now that there is a textbook for this course, I will be able to hand it off to someone else to teach. I thank all of my colleagues and students (too numerous to mention by name, for fear of omitting someone) who suggested interesting questions or innovative solutions. I thank our editor at Princeton, Vickie Kearn, who encouraged us to write the book and helped us throughout the writing process. In any successful collaboration, both parties have to be willing to do 90% of the work. I thank my coauthor, John, for definitely doing more than his share and for making the project fun. I am also grateful to my wife and children, Carol, Lee, and Rachel, for their specific suggestions and corrections, for their general support of this project, and for making life fun and worthwhile.

Preface

How big IS it?

Numbers are thrown at us all the time. They are frequently used to scare us: "Shark attacks doubled this year!" or "Dozens of lives could be saved by using infant car seats on airplanes!" They are often used to tempt us: "This week's lottery prize is $100 million!" They are certainly needed to understand the world around us: "The average American produces 100 cubic feet of garbage every year!" or "Nuclear power plants produce tons of high-level radioactive waste!"

You can make sense of these often confusing and sometimes contradictory numbers with just two tools: (1) an understanding of the meaning of large numbers and (2) an ability to make rough, common-sense, estimates starting from just a few basic facts. We'll teach you these straightforward skills so you can better understand the world around you and better recognize numerical, political, and scientific nonsense.

You can also use these tools to further your career. Many top companies use estimation questions in job interviews to judge the intelligence and flexibility of their applicants [1]. Leading software firms, management consultants, and investment banks (for example, Microsoft, Goldman Sachs, and Smith Barney) ask questions such as What's the size of the market for disposable diapers in China? How many golf balls does it take to fill a 747? and How many piano tuners are there in the world? [2, 3] Companies use these questions as an excellent test of the applicants' abilities to think on their feet and to apply their mathematical skills to real-world problems.

These problems are frequently called "Fermi problems," after the legendary physicist Enrico Fermi, who delighted in creating and solving them. During one of the first atomic bomb tests, Fermi supposedly dropped a few scraps of paper as the shock wave passed and estimated the strength of the blast from the motion of the scraps as they fell.

In this book, we will help you develop the ability to estimate almost anything, from the amount of landfill space needed to the number of people in the world picking their nose at this instant. As there is no single correct way to analyze these questions, we will indicate some of the many paths to the right answer.

We will start with two short chapters on how to estimate and how to handle large numbers, and then we'll move on to the heart of the book: interesting questions (with lots of hints if you want them) followed by the answers on the other side of the page. The questions are divided into chapters, each focusing on a particular topic, such as energy and the environment, transportation, and risk. Each chapter will start with easier questions and work up to harder ones. The questions in chapters 6 through 9 will cover energy in its various forms. We will start with mountain climbing and go on to compare gasoline, chocolate chip cookies, batteries, the Sun, gerbils, windmills, and uranium.

The questions cover a large range of phenomena, from the simple to the complex and from the silly to the serious. We will figure out the answers to many fascinating questions, including the following:

- If all the people in the world were crammed together, how much space would we occupy?
- How many batteries would it take to replace your car's gas tank?
- Could Spider-Man really stop a subway car?
- How much waste do nuclear and coal electric power generators make per year?

- What does it really cost to drive a car?
- Which is more powerful per pound, the Sun or a gerbil?
- How much more cropland would we need to re-place gasoline with corn-based ethanol?

All you will need to answer these questions is a willingness to think and to handle big numbers. We will remind you of the few scientific principles and equations you might need. You will be astonished at how much you can figure out starting with the knowledge you already have.

The new knowledge you will attain can be applied to all other estimation problems you may come across in the future. Oh, and good luck on that job interview!

gue??timation

* * * * * * * * * * * * * * * * *

How to Solve Problems

Chapter 1

* * * * * * * * * * * * * * * *

STEP 1: Write down the answer [4]. In other words, come up with a reasonably close solution. This is frequently all the information you need.

For example, if it is 250 miles from New York to Boston, how long will it take to drive? You would immediately estimate that it should take about four or five hours, based on an average speed of 50–60 mph. This is enough information to decide whether or not you will drive to Boston for the weekend. If you do decide to drive, you will look at maps or the Internet and figure out the exact route and the exact expected driving time.

Similarly, before you go into a store, you usually know how much you are willing to spend. You might think it is reasonable to spend about $100 on an X-Game2. If you see it for $30, you will automatically buy it. If it sells for $300, you will automatically not buy it. Only if the price is around $100 will you have to think about whether to buy it.

We will apply the same reasoning here. We'll try to estimate the answer to within a factor of ten. Why a factor of ten? Because that is good enough to make most decisions.

Once you have estimated the answer to a problem, the answer will fall into one of the three "Goldilocks" categories:

1. too big

2. too small

3. just right

If the answer is too big or too small, then you know what to do (e.g., buy the item, don't drive to Boston). Only if the answer is just right will you need to put more work into solving the problem and refining the answer. (But that's beyond the scope of this book. We just aim to help you estimate the answer to within a factor of ten.)

If all problems were as simple as that, you wouldn't need this book. Many problems are too complicated for you to come up with an immediate correct answer. These problems will need to be broken down into smaller and smaller pieces. Eventually, the pieces will be small enough and simple enough that you can estimate an answer for each one. And so we come to

STEP 2: If you can't estimate the answer, break the problem into smaller pieces and estimate the answer for each one. You only need to estimate each answer to within a factor of ten. How hard can that be?

It is often easier to establish lower and upper bounds for a quantity than to estimate it directly. If we are trying to estimate, for example, how many circus clowns can fit into a Volkswagen Beetle, we know the answer must be more than one and less than 100. We could average the upper and lower bounds and use 50 for our estimate. This is not the best choice because it is a factor of 50 greater than our lower bound and only a factor of two lower than our upper bound.

Since we want our estimate to be the same factor away from our upper and lower bounds, we will use the geometric mean. To take the *approximate* geometric mean of any two numbers, just average their coefficients and average their exponents.* In the clown case, the geometric mean of one $(10^0)^†$ and 100 (10^2) is 10 (10^1) because one is the average of the exponents zero and two. Similarly, the geometric mean of 2×10^{15}

---

* We use coefficients and exponents to describe numbers in scientific notation. The exponent is the power of ten and the coefficient is the number (between 1 and 9.99) that multiplies the power of ten. If you are not familiar with this notation, please quickly check the section on scientific notation ("Dealing with Large Numbers") and then come right back. We'll wait for you here.

† Any number raised to the 0th power is 1.

$1 = 10^0$
$100 = 10^2$ geometric mean $= 10$

and 6×10^3 is about 4×10^9 (because $4 = \frac{2+6}{2}$ and $9 = \frac{15+3}{2}$).* If the sum of the exponents is odd, it is a little more complicated. Then you should decrease the exponent sum by one so it is even, and multiply the final answer by three. Therefore, the geometric mean of one and 10^3 is $3 \times 10^1 = 30$.

EXAMPLE 1: MongaMillions Lottery Ticket Stack

Here's a relatively straightforward example: Your chance of winning the MongaMillions lottery is one in 100 million.[†] If you stacked up all the possible different lottery tickets, how tall would this stack be? Which distance is this closest to: a tall building (100 m or 300 ft), a small mountain (1000 m), Mt Everest (10,000 m), the height of the atmosphere (10^5 m), the distance from New York to Chicago (10^6 m), the diameter of the Earth (10^7 m), or the distance to the moon (4×10^8 m)? Imagine trying to pick the single winning ticket from a stack this high.

Solution: To solve this problem, we need two pieces of information: the number of possible tickets and the thickness of each ticket. Because your chance of winning is one in 100 million, this means that there are 100 million (10^8) possible different tickets.[‡] We can't reliably estimate really thin items like a single lottery ticket (is it 1/16 in. or 1/64 in.? is it 1 mm or 0.1 mm?) so let's try to get the thickness of a pack of tickets.

---

* To be more precise (which this book rarely is), the geometric mean of two numbers, b and c, is $a = \sqrt{bc}$. Our approximate rule is exact for the exponents and close enough for this book for the coefficients.

† Lottery billboards frequently have the odds of winning in very small print at the bottom.

‡ 100 million = 100,000,000 or 1 followed by eight (count them!) zeros. This can be written in scientific notation as 1×10^8.

Let's think about packs of paper in general. One ream of copier or printer paper (500 sheets) is about 1.5 to 2 in. (or about 5 cm since 1 in. = 2.5 cm) but paper is thinner than lottery tickets. A pack of 52 playing cards is also about 1 cm. That's probably closer. This means that the thickness of one ticket is

$$t = \frac{1 \, \text{cm}}{52 \, \text{tickets}} = 0.02 \frac{\text{cm}}{\text{ticket}} \times \frac{1 \, \text{m}}{10^2 \, \text{cm}}$$

$$= 2 \times 10^{-4} \frac{\text{m}}{\text{ticket}}$$

Therefore, the thickness of 10^8 tickets is

$$T = 2 \times 10^{-4} \frac{\text{m}}{\text{ticket}} \times 10^8 \, \text{tickets} = 2 \times 10^4 \, \text{m}$$

2×10^4 m is 20 kilometers or 20 km (which is about 15 miles since 1 mi = 1.6 km).

If stacked horizontally, it would take you four or five hours to walk that far.

If stacked vertically, it would be twice as high as Mt Everest (30,000 ft or 10 km) and twice as high as jumbo jets fly.

Now perhaps you used the thickness of regular paper so your stack is a few times shorter. Perhaps you used 1 mm per ticket so your stack is a few times taller. Does it really matter whether the stack is 10 km or 50 km? Either way, your chance of pulling the single winning ticket from that stack is pretty darn small.

EXAMPLE 2: Flighty Americans

These problems are great fun because, first, we are not looking for an exact answer, and second, there are many different ways of estimating the answer. Here is a slightly harder question with multiple solutions.

How many airplane flights do Americans take in one year?

We can estimate this from the top down or from the bottom up. We can start with the number of airports or with the number of Americans.

Solution 1: Start with the number of Americans and estimate how many plane flights each of us take per year. There are 3×10^8 Americans.* Most of us probably travel once a year (i.e., two flights) on vacation or business and a small fraction of us (say 10%) travel much more than that. This means that the number of flights per person per year is between two and four (so we'll use three). Therefore, the total number of flights per year is

$$N = 3 \times 10^8 \text{ people} \times 3 \text{ flights/person-year}$$

$$= 9 \times 10^8 \text{ passengers/year}$$

Solution 2: Start with the number of airports and then estimate the flights per airport and the passengers per flight. There are several reasonable size airports in a medium-sized state (e.g., Virginia has Dulles, Reagan-National, Norfolk, Richmond, and Charlottesville; and Massachusetts has Boston and Springfield). If each of the fifty states has three airports then there are 150 airports in the US. Each airport can handle at most one flight every two minutes, which is 30 flights per hour or 500 flights per 16-hour day. Most airports will have many fewer flights than the maximum. Each airplane can hold between 50 and 250 passengers. This means

---

* This is one of those numbers you need to know to do many estimation questions. Go ahead and write it on your palm so you'll be prepared for the test at the end of the book.

that we have about

$$N = 150 \text{ airports} \times \frac{100 \text{ flights}}{\text{airport-day}} \times \frac{100 \text{ passengers}}{\text{flight}}$$

$$\times \frac{365 \text{ days}}{\text{year}} = 5 \times 10^8 \text{ passengers per year}$$

Wow! Both methods agree within a factor of two.

The actual number of US domestic airline passengers in 2005 was 6.6×10^8, which is close enough to both answers.

EXAMPLE 3: Piano Tuners in Los Angeles

Now let's work out a harder problem.

How many piano tuners are there in Los Angeles (or New York or Virginia Beach or your own city)? This is the classic example originated by Enrico Fermi [5] and used at the beginning of many physics courses because it requires employing the methods and reasoning used to attack these problems but does not need any physics concepts.

Solution: This is a sufficiently complicated problem that we cannot just estimate the answer. To solve this, we need to break down the problem. We need to estimate (1) how many pianos there are in Los Angeles and (2) how many pianos each tuner can care for. To estimate the number of pianos, we need (1) the population of the city, (2) the proportion of people that own a piano, and (3) the number of schools, churches, etc. that also have pianos. To estimate the number of pianos each tuner can care for, we need to estimate (1) how often each piano is tuned, (2) how much time it takes to tune a piano, and (3) how much time a piano tuner spends tuning pianos.

This means that we need to estimate the following:

1. population of Los Angeles
2. proportion of pianos per person
3. how often each piano is tuned per year
4. how much time it takes to tune each piano
5. how much time each piano tuner works per year

Let's take it from the top.

1. The population of Los Angeles must be much less than 10^8 (since the population of the US is 3×10^8). It must be much more than 10^6 (since that is the size of an ordinary big city). We'll estimate it at 10^7.

2. Pianos will be owned by individuals, schools, and houses of worship. About 10% of the population plays a musical instrument (it's surely more than 1% and less than 100%). At most 10% of musicians play the piano and not all of them own a piano so the proportion that own a piano is probably 2–3% of the musicians. This would be 2×10^{-3} of the population. There is about one house of worship per thousand people and each of those will have a piano. There is about one school per 500 students (or about 1 per 1000 population) and each of those will have a piano. This gives us about 4 or 5 $\times 10^{-3}$ pianos per person. Thus, the number of pianos will be about $10^7 \times 4 \times 10^{-3} = 4 \times 10^4$.

3. Pianos will be tuned less than once per month and more than once per decade. We'll estimate once per year.

4. It must take much more than 30 minutes and less than one day to tune a piano (assuming that it is not too badly out of tune). We'll estimate

2 hours. Another way to look at it is that there are 88 keys. At 1 minute per key, it will take 1.5 hours. At 2 minutes per key, it will take 3 hours.

5. A full-time worker works 8 hours per day, 5 days per week, and 50 weeks per year which gives $8 \times 5 \times 50 = 2000$ hours. In 2000 hours she can tune about 1000 pianos (wow!).

This means that the 4×10^4 pianos need 40 piano tuners.

How close are we? Well, the Yellow Pages for our city of 10^6 inhabitants (ten times fewer than LA) has 16 entries under the heading of "Pianos—Tuning, Repairing & Refinishing." There are probably only one or two tuners per entry and they probably do not spend full time tuning. This means that our estimate is probably too low by a factor of five. However, that is a LOT closer than we could get by just guessing.

Remember that we are only trying to estimate the answer within a factor of ten.

Dealing with Large Numbers

Chapter 2

* * * * * * * * * * * * * * * *

2.1 Scientific Notation

As you may have noticed, we used 10^8 instead of 100,000,000 to represent 100 million. We do this for two reasons. The first is that, if we multiply 3 trillion times 20 quadrillion like this:

$$3000000000000 \times 20000000000000000$$

$$= 6000000\ldots\ldots$$

we will almost certainly miscount all those pesky zeros. If we use a calculator, we will first miscount the pesky zeros and then we will enter the incorrect number of zeros in the calculator so the number will be even more wrong. We'll get an answer with the correct first digit (6) but the incorrect size. That's like getting \$60 when you should have received \$6000. The number of zeros is much more important than the leading digit.

There is an easy and compact way of writing very large and very small numbers. Any number can be written as the product of a number between 1 and 10 and a number that is a power of ten. For example, we can write 257 as 2.57×100 and 0.00257 as 2.57×0.001. Now we have to count the zeros (but only once per number). One hundred (100) has two zeros so we write it as 10^2, and 0.001 has three zeros (counting the leading zero) and is less than one so we write it as 10^{-3}. Therefore, we write 257 as 2.57×10^2 and 0.00257 as 2.57×10^{-3}. The *exponent* tells us how many zeros are in the power of ten (2 and -3 in the previous sentence) and the *coefficient* multiplies the power of ten. This system is called *scientific notation*.

Here are some examples to make this clearer:

$$0.01 = 1 \times 10^{-2}$$
$$2000 = 2 \times 10^3$$
$$3,000,000 = 3 \times 10^6$$

The second reason for using scientific notation of the form $x \times 10^y$ is that the most important part of the number is the exponent y and not the coefficient x. When we write the 300 million population of the United States as 3×10^8, the 8 is much more important than the 3. This is because if you change the 3 to a 4, you change the population by only 1/3 or 30%. If you change the 8 to a 9, then you change the population by a factor of 10 (or 1000%). That is a huge change, especially if you think that we are already too crowded here. Thus, we use scientific notation so that the exponent is written explicitly.

The rules for multiplying and dividing with scientific notation are straightforward. When we multiply numbers, we multiply coefficients and add exponents. For example,

$$3 \times 10^6 \times 4 \times 10^8 = (3 \times 4) \times 10^{6+8}$$
$$= 12 \times 10^{14} = 1.2 \times 10^{15}$$

When we divide numbers, we divide coefficients and subtract exponents. For example,

$$\frac{3 \times 10^6}{4 \times 10^8} = \frac{3}{4} \times 10^{6-8} = 0.75 \times 10^{-2} = 7.5 \times 10^{-3}$$

Note that in both of these examples, we had to cope with a coefficient either more than 10 or less than 1. In these cases, we will rewrite the out-of-range coefficient in scientific notation itself. Thus, in the first example the coefficient 12 can be written as 1.2×10^1. This means that what we really did was

$$12 \times 10^{14} = (1.2 \times 10^1) \times 10^{14} = 1.2 \times 10^{15}$$

In the second case, the coefficient 0.75 is rewritten as 7.5×10^{-1} so that

$$0.75 \times 10^{-2} = (7.5 \times 10^{-1}) \times 10^{-2} = 7.5 \times 10^{-3}$$

When we add or subtract numbers using scientific notation, the exponents of both numbers must be the same. To add 3×10^7 and 4×10^8, we need to convert the number with the smaller exponent into a form where it has the same exponent as the other number. In this case, when we increase the exponent by one (from 7 to 8), we must simultaneously divide the coefficient by 10 (this is because increasing the exponent increases the number by a factor of ten so we must decrease the coefficient to compensate and keep the number the same). Thus, we have

$$3 \times 10^7 + 4 \times 10^8 = 0.3 \times 10^8 + 4 \times 10^8 = 4.3 \times 10^8$$

Going back to that first equation, we write 3 trillion (3000000000000) as 3×10^{12} and 20 quadrillion (20000000000000000) as 2×10^{16} so that the operation becomes

$$3 \times 10^{12} \times 2 \times 10^{16} = (2 \times 3) \times 10^{12+16} = 6 \times 10^{28}$$

We no longer need to count zeros; all we have to do is to add exponents. We have a much easier time adding 12 and 16 to get 28 than we do counting 12 zeros and 16 zeros and then writing down 28 zeros.

2.2 Accuracy

The most important part of any number is the exponent. After that, the next most important number is the first digit of the coefficient (the number that multiplies the power of ten). The second and subsequent digits of the coefficient are just small corrections to the first digit.

The number of digits in the coefficient (also called the "number of significant figures") tells us how well we know that number. For example, if your friend gives you driving directions, there is a tremendous difference between the direction to "drive a couple of dozen miles east and then turn left on Obscure Alley" and the direction to "drive 25.2 miles east and then turn left on Obscure Alley." The first direction is rather vague and imprecise. You expect to find Obscure Alley somewhere between 12 and 36 miles away. If you miss the left turn, you will probably drive quite far before you turn around to look for it again. The second direction is very precise and you expect to find Obscure Alley between 25.1 and 25.3 miles away. If you miss the turn, you will probably turn around by the time you have driven 26 miles. The extra digits in the second set of directions imply that your friend has measured the distance carefully.

Similarly, the silliness of having too many digits is illustrated by the following anecdote. Suppose that you ask a museum guard how old a dinosaur skeleton is. He responds that it is 75 million and 3 (75,000,003) years old. When you look puzzled, he explains that when he started the job three years ago, the skeleton was already 75 million years old.

Many of us make the same kind of mistake with our calculators. Suppose that we used 23.0 gallons of gasoline to drive 327 miles. If we divide 327 by 23 on our calculator we get 14.2173913. . .. But this cannot be the answer to our problem. We did not measure either the miles driven or the fuel consumed to 1 part in a billion so our answer cannot possibly be so precise. Our gas mileage should be 327 mil/23.0 gal = 14.2 mpg.

There are lots of rules for how to deal with significant figures in scientific calculations. Fortunately, we will not need most of them.

In this book, we will estimate quantities to within a factor of ten. Therefore, we will almost always keep only one digit in the coefficient. This means that we will round off 7.2×10^3 to just 7×10^3. Our estimates are just not accurate past the first digit. Keeping more digits is lying; it claims we know the answer much better than we actually do.

There is another benefit to keeping only one significant figure: you shouldn't need a calculator to solve these problems. If you can add or subtract one- or two-digit numbers (the exponents) without a calculator and multiply or divide one-digit numbers without a calculator, then you are all set to proceed.

2.3 A Note on Units

We have been using the metric system as students and scientists for decades now, but we still think in inches, feet, pounds, and degrees Fahrenheit (US customary units) because those are the units we use every day. When we estimate things, we'll typically estimate them in US units and then convert to metric. When we're done with the problem, we might convert back to US units. If we used metric units in daily life, we would have to do this for some quantities anyway, since units such as km/hr, liters (L), and cubic centimeters (cm^3) are not part of the MKS (meter/kilogram/second) system.

Why would anyone bother to do all that? Well, it's a lot easier to do the calculations in metric because the conversion factors are (1) merely powers of ten (*you* try converting miles to inches without a calculator), and (2) converting quantities like volume is also much easier in the metric system. We know that $1\,L = 1000\,cm^3$ and $1\,m^3 = 1000\,L$. What is a gallon in terms of cubic inches or a cubic foot in terms of cups? We sure don't know.

In addition to simplifying unit conversions, the metric system (SI or International System of Units) is much easier for doing many calculations. All the quantities are standardized on the meter, second,

Quantity	Metric Unit	US Customary equivalent
Length	1 meter (m)	3 ft
Length	10^3 m (1 km)	0.6 mi
Length	0.01 m (1 cm)	0.4 in.
Volume	1 liter (L)	1 quart
Mass	1 kilogram (kg)	Mass of 2.2 pounds (lb)
Mass	10^3 kg	1 ton
Weight	1 newton (N)	0.2 lb
Speed	1 m/s	2.2 mph
Time	$\pi \times 10^7$ s	1 year

and kilogram. Thus, the unit of force, the newton (N), is equal to a kilogram–meter/second². We will use the following abbreviations for units: meter (m), second (s), kilogram (kg), watt (W), joulc (J), newton (N), liter (L), hour (hr), etc. For more information on SI units, see the US National Institute of Standards and Technology (NIST) website: http://physics.nist.gov/cuu/Units/index.html. We will use the following abbreviations for US customary units: mi (milc), ft (foot), in. (inch), gal (gallon), and lb (pound).

We will also use the standard prefixes giga or 10^9 (G), mega or 10^6 (M), kilo or 10^3 (k), centi or 10^{-2} (c), milli or 10^{-3} (m), micro or 10^{-6} (μ), and nano or 10^{-9} (n). These are tabulated in the appendix. If we need to use pico, tera, or yocto, we'll warn you first.

2.4 Unit Conversion

We will frequently need to convert a quantity from one unit to another. For example, to calculate the distance light travels in one year or the energy used by a 100-W light bulb in one year, you need to convert time from 1 year to some number of seconds. To do this, we will multiply our original number by various conversion factors that individually are equal to one, e.g., $\frac{60\,\text{s}}{1\,\text{min}} = 1$. Thus,

$$1\,\text{year} = 1\,\text{year} \times \left(\frac{365\,\text{days}}{1\,\text{year}}\right)\left(\frac{24\,\text{hr}}{1\,\text{day}}\right)$$

$$\times \left(\frac{60\,\text{min}}{1\,\text{hr}}\right)\left(\frac{60\,\text{s}}{1\,\text{min}}\right) = 3.15 \times 10^7\,\text{s}$$

Note that 1 year $\approx \pi \times 10^7$ s. π appears in this because the Earth goes around the Sun in an almost perfect circle and the circumference of a circle $c = 2\pi R^*$.

Another handy conversion is from miles per hour (mph) to meters per second (m/s). We think of speeds in mph, but it is much easier to do calculations in m/s. Fortunately, the conversion is straightforward. We first convert meters to miles via kilometers and then we convert seconds to hours via minutes:

$$1\text{m/s} = 1\text{m/s} \times \left(\frac{1\text{km}}{10^3\,\text{m}}\right)\left(\frac{0.6\,\text{mi}}{1\,\text{km}}\right)$$

$$\times \left(\frac{60\,\text{s}}{1\,\text{min}}\right)\left(\frac{60\,\text{min}}{1\,\text{h}}\right) = 2.2\,\text{mph}$$

Thus, 1 m/s is a little more than 2 mph.

We'll use both these handy facts later on. Feel free to write them on your palm also.

---

* You didn't really believe that, did you? The π is just a coincidence, but it makes a handy mnemonic. Of course, the exponent "7" is much more important than the leading digit "3".

General Questions

Let's start with some straightforward questions about distance and space. We'll ask about how much space we need for ourselves, how much space we need for our garbage, and how much space we need for our pickles.

Chapter 3

One big family

3.1

If all the humans in the world were crammed together, how much area would we require? Compare this to the area of a large city, a state or small country, the US, Asia.

How much area would we need if we gave every family a house and a yard (i.e., a small plot of land)?

✳ ✳ ✳ ✳ ✳ ✳ ✳ ✳ ✳ ✳ ✳ ✳ ✳ ✳ ✳

HINT: The population of the world is a tad over 6 billion. The tad in this case is a mere 300 million (0.3 billion) at the time of writing. We'll ignore it.

HINT: How many people can fit into a square meter? Recall that a meter is slightly longer than a yard, so a square meter is about 10 square feet.

HINT: Once you've decided on the area you want for each person, then multiply that by the number of people.

HINT: To estimate the area of a yard, assume it to be square. How wide is your yard? 10 ft (3 m), 30 ft (10 m), 100 ft (30 m), 300 ft (100 m)? If you chose 10 m for the width, then the area of your yard is $A = 10\text{ m} \times 10\text{ m} = 100\text{ m}^2$.

21

ANSWER: Okay, 6 billion people is 6×10^9 of us. How many can we cram into a square meter (i.e., 3 ft by 3 ft)? We're not sure, but it is certainly between 3 and 10. We'll choose 6. (That ignores the space needed to play, eat, sleep, and well, the porta-potty question is in chapter 4, so can you wait until then?) If there are 6 people per square meter, then 6 billion people will need

$$A = 6 \times 10^9 \text{people} \times \frac{1 \, \text{m}^2}{6 \, \text{people}} = 10^9 \, \text{m}^2$$

We have no idea how big a billion square meters is (although it sure sounds like a lot) so let's convert it to more reasonable units. We'll convert it to kilometers (km) and to miles. $1 \, \text{km} = 10^3$ m. A square kilometer is a square with sides of 10^3 m so that $1 \, \text{km}^2 = 10^3 \, \text{m} \times 10^3 \, \text{m} = 10^6 \, \text{m}^2$.

$$A = 10^9 \, \text{m}^2 \times \frac{1 \, \text{km}^2}{10^6 \, \text{m}^2} = 10^3 \, \text{km}^2$$

(Just to remind you, when you divide numbers in scientific notation, you divide the coefficients and subtract the exponents. In this case, $10^9/10^6 = 10^{9-6} = 10^3$.) Thus, we would occupy an area of 1000 square kilometers. That's a square 30 km on a side. Since a mile is about 1.5 km, all the people on Earth would fit in a square that is 30 km or 20 mi on a side. That's the area of a large city such as Los Angeles or Virginia Beach.

Wow! That's not much at all.

Now let's give every family a house and a yard (a small piece of land). First we need to estimate the size of the average family. In the US and Europe, the average family is about three people, but in the developing world, it's a bit more than that. We'll choose three so that we overestimate the amount of land.

Next, we need to estimate the size of the yard. Since we are not good at estimating area, we'll assume a square yard and estimate its width. It will definitely be smaller than a football field (100 m or 300 ft) and bigger than a house (10 m or 30 ft) so we'll take the geometric mean and estimate 30 m. This means that each family gets a piece of land of area $A = 30\,\text{m} \times 30\,\text{m} = 10^3\,\text{m}^2$ (or about 1/4 acre). Therefore, all of us together will use a total land area of

$$A = 6 \times 10^9 \text{people} \times \frac{1\,\text{family}}{3\,\text{people}} \times \frac{10^3\,\text{m}^2}{\text{family}} = 2 \times 10^{12}\,\text{m}^2$$

(Just to remind you, when you multiply numbers in scientific notation, you multiply the coefficients and add the exponents. In this case, $10^9 \times 10^3 = 10^{9+3} = 10^{12}$.) This is $2 \times 10^6\,\text{km}^2$ (two million square kilometers) or about $1 \times 10^6\,\text{mi}^2$ (one million square miles). While it sounds like a lot, it is only the area of Alaska or twice the area of Egypt. That is only 1% of the surface area of the Earth.

That could leave a lot of room for other species!

Fore!

3.2

How many golf balls would it take to circle the Earth at the equator?*

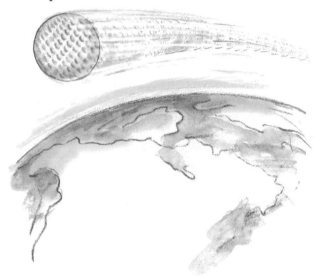

* Thanks to Tom Isenhour for this question [6]. Can we have that raise now, Tom?

* * * * * * * * * * * * * * *

HINT: What's the diameter of a golf ball? Remember that 1 inch is 2.5 cm.

HINT: What is the circumference of the Earth?

HINT: If you remember then the circumference is $c = 2\pi R$. (If you don't remember the radius, c is still $2\pi R$ but the formula is a lot less useful.)

HINT: There is a 3-hour time difference between LA and NY. There are 24 time zones total.

HINT: The Earth's circumference is eight times the distance from NY to LA. You can fly that distance in 6 hours.

ANSWER: To answer this question we need the diameter of a golf ball and the circumference of the Earth. Let's start with the easier part. A golf ball has a diameter of a bit less than 2 inches or about 4 cm.

There are several ways to estimate the circumference of the Earth. For example, there is a three-hour time difference between New York and Los Angeles and there are 24 times zones covering the Earth. Therefore, the circumference of the Earth is about eight times the distance from NY to LA. If you don't remember that the distance between them is 3000 miles, then you can estimate it from the fact that it takes about six hours to fly from NY to LA and a modern jet flies at about 500 mph. Thus, the circumference is about $c = 8 \times 3000\,\text{mi} = 2.4 \times 10^4\,\text{mi}$.

Alternatively, we know that passenger jets fly slower than the Earth rotates (since you always arrive after you leave [in local time]) and that some military jets can fly faster than the Earth rotates. Since passenger jets fly at about 500 mph and military jets can fly up to 2000 mph, we can estimate the Earth's rotation at 1000 mph. Since the Earth rotates completely in 24 hours, its circumference must be $c = 24 \times 1000\,\text{mph} = 2.4 \times 10^4\,\text{mi}$.

Of course, if you remembered that the circumference is 25,000 miles (or 40,000 km) or that the radius of the Earth is 4000 miles (6400 km) and the circumference is $c = 2\pi R$, then you didn't need to estimate it.

Now the arithmetic is straightforward. First we need to convert the circumference of the Earth from kilometers to centimeters. The number of golf balls needed is

$$N = 4 \times 10^4\,\text{km} \times \frac{10^3\,\text{m}}{1\,\text{km}} \times \frac{10^2\,\text{cm}}{1\,\text{m}} \times \frac{1\,\text{golf ball}}{4\,\text{cm}}$$

$$= 10^9 \text{golf balls}$$

The Pacific Ocean is a very large water hazard. It would be extremely irritating to lose a billion golf balls in the water! We'd better use the special kind that floats.

This also provides an interesting peg on which to hang the concept of "parts per billion" (ppb). If the air contains so many ppb of some potentially toxic substance, that is about the number of, say, red golf balls in the otherwise white ones surrounding the Earth. You could walk along the equator for months before finding your first red golf ball.

This is a fine pickle
you've got us into, Patty

If all the pickles sold in the US last year were
placed end-to-end, what distance
would they cover?

HINT: How many pickles does the average American eat in
a year?

HINT: How long is a typical pickle?

ANSWER: We will need to estimate the number of pickles the average American eats each year and the length of the average pickle. Our average pickle-in-the-street is definitely more than 1 cm and less than 100 cm (a 1-m pickle? yikes!) so we'll estimate that it is about 10 cm (4 in.) long. The average American consumes more than one pickle per year (that's including the pickle slices on hamburgers) and less than one per day (400 per year) so we'll estimate 20 pickles per year.* Therefore, the total length of all the pickles consumed in one year is

$$L = 3 \times 10^8 \text{ people} \times \frac{20 \text{ pickles}}{\text{person}} \times \frac{10 \text{ cm}}{\text{pickle}}$$

$$= 6 \times 10^{10} \text{ cm} \times \frac{1 \text{ m}}{10^2 \text{ cm}}$$

$$= 6 \times 10^8 \text{ m}$$

That is a distance of 6×10^5 kilometers or 4×10^5 miles. That is more than the distance from the Earth to the Moon! Who needs advanced technology? NASA could just take the space pickle.

---

* According to the USDA [7], Americans consumed about four pounds of pickles per person (try saying that five times fast!) in 2000 so our estimate of 20 pickles per person is not too bad.

Throwing in the towel

3·4

What is the surface area of a typical bath towel (include the fibers!). Compare this to the area of a room, a house, a football field.

* * * * * * * * * * * * * * *

HINT: Think about the little fibers in a really fluffy towel; how many are there in a square centimeter or square inch?

HINT: What is the area of a large towel? How many total fibers does it have?

HINT: What is the surface area of each little fiber?

HINT: Consider the length and thickness of each fiber.

ANSWER: That's obvious, surely! A large rectangular towel 1 m by 2 m has a total surface area of 4 m² (including both sides), right? (In US units, a big towel may be as large as 3 ft by 6 ft.)

Wrong, actually, unless it is a very worn-out towel. New towels have many little fibers that can absorb a lot of moisture (recall the old puzzle—what gets wetter the more it dries?). Unless you're a fan of the *Hitchhiker's Guide to the Galaxy*, you won't have brought your own towel, so nip off to the bathroom and examine one; quickly now, we're dripping all over the floor.

You don't need to actually go and count the number of fibers per square inch or per square centimeter; in the latter case there must be more than 10 and fewer than 1000, so we take the geometric mean of 10^1 and 10^3, which is 10^2. In a square inch, being about 6 cm², we should expect about six times as many. This will of course vary, depending on where you buy your towels; we are assuming that we are describing one of those very nice towels found in one of those very nice hotels.

Back already? Right-oh. Now we need to estimate the surface area of each fiber. We can approximate the fiber as a cylinder or a box. Cylinders are complicated so we'll use boxes. Each fiber is about 0.5 cm (1/4 in.) long and 1 mm (0.1 cm) wide. Each "boxy" fiber then has four flat surfaces, each 0.5 cm by 0.1 cm. Thus, the surface area of one fiber is

$$A_{\text{fiber}} = 4 \times 0.5 \, \text{cm} \times \frac{1 \, \text{m}}{10^2 \, \text{cm}}$$

$$\times 0.1 \, \text{cm} \times \frac{1 \, \text{m}}{10^2 \, \text{cm}}$$

$$= 2 \times 10^{-5} \, \text{m}^2$$

Now we can calculate the total surface area of our big bathroom towel:

$$A_{\text{total}} = \text{towel area} \times \text{fibers per area} \times \text{area per fiber}$$

$$= 4\,\text{m}^2 \times \frac{10^2 \text{ fibers}}{\text{cm}^2} \times \frac{10^4 \text{ cm}^2}{1\,\text{m}^2} \times \frac{2 \times 10^{-5}\,\text{m}^2}{\text{fiber}}$$

$$= 80\,\text{m}^2$$

That is about 800 square feet: the size of a large apartment or a small house.

This problem is similar in some ways to calculating the length of the coastline. Just as the area of the towel is much greater than its simple area, the length of coast from, say, New York to Boston is much more than the 200-mile driving distance.

How long would it
take a running
water faucet* to
fill the (inverted)
dome of the US
Capitol building
or St. Paul's Cathedral?
Give your answer in
seconds, days, weeks,
or whatever units seem
reasonable.

*That's a "tap" for our British readers (and author).

✳ ✳ ✳ ✳ ✳ ✳ ✳ ✳ ✳ ✳ ✳ ✳ ✳ ✳ ✳

HINT: Estimate the diameter of the dome.

HINT: Volume $\approx (1/4)d^3$.

HINT: How long does it take you to fill a one gallon (4 liter) jug
of water from your kitchen faucet? Alternatively, what is the flow
rate in gallons per minute of a modern showerhead?

HINT: There are 4 L/gal and 10^3 L/m³.

ANSWER: We need to estimate the volume of the dome and the flow rate of a water faucet. To estimate the volume of the dome, we need to estimate its diameter. The diameter of the dome of St. Paul's or the Capitol Building is more than 10 m (or 30 ft) and less than 100 m (the length of a football field, about 300 ft), so we take the geometric mean and estimate it as $\sqrt{10 \times 100}$ m = 30 m (100 ft) in diameter.

If we remember that the volume of a sphere is $V = \frac{4}{3}\pi R^3$ and that a dome is half a sphere, then we have

$$V = \frac{1}{2}\frac{4}{3}\pi R^3 = 2(15\text{m})^3 = 6 \times 10^3\text{m}^3$$

If we forgot the equation of a sphere, then we could pretend that the dome is half of a cube (as Picasso might have done) and approximate the volume as

$$V = \frac{1}{2}d^3 = 0.5 \times (30\,\text{m})^3 = 10^4\text{m}^3$$

We would be off by only a factor of two. Not a problem!

Now we need to estimate the flow rate of a water faucet. A typical domestic faucet running full-tilt can fill a one-gallon container in under 30 seconds. A US low-flow shower is limited to 2.5 gallons per minute so that is about the same. A cubic meter of water is 10^3 liters, or 250 gallons. Thus, the time to fill the Capitol dome is

$$t = \frac{\text{volume of dome}}{\text{flow rate}}$$

$$= \frac{6 \times 10^3\,\text{m}^3 \times 2.5 \times 10^2\,\text{gal/m}^3}{2\,\text{gal/min}}$$

$$= 7 \times 10^5\,\text{min}$$

Seven hundred thousand minutes is not a very helpful quantity. Let's convert it to more useful units and see what we get.

There are 60 minutes in an hour, or about $60 \times 25 = 1500$ minutes in a day.* We can convert minutes to days to get

$$t = 7 \times 10^5 \, \text{min} \times \frac{1 \, \text{day}}{1.5 \times 10^3 \, \text{min}} = 500 \, \text{days}$$

That is less than two years.

Of course, we'd have to invert the dome first...

---

* We often wish we had an extra hour in the day. In this case, it's just for easier calculation.

A mole of cats

How massive is a mole of cats?* (A mole is the number of atoms that weigh that element's atomic weight in grams. For example, a mole of hydrogen weighs 1 gram and a mole of carbon weighs 12 grams. It is used in chemistry to make sure that there are equivalent numbers of atoms for a chemical reaction.) Compare this to the mass of a mountain, a continent, the moon (7×10^{22} kg), the Earth (6×10^{24} kg).

* Thanks again, Tom [8].

✳ ✳ ✳ ✳ ✳ ✳ ✳ ✳ ✳ ✳ ✳ ✳ ✳ ✳ ✳

HINT: What is the weight of a typical adult domestic cat?

HINT: Remember that one mole of anything contains Avogadro's number (6×10^{23}) of those items.

ANSWER: We'll use fat cats weighing about 8 kg (18 lb) each. We're using a neighbor's cat, Quentin, as a model here, though without his permission (or the neighbor's). There are $N_A = 6 \times 10^{23}$ items in a mole, whether you are talking about a mole of atoms or a mole of cats (or a mole of moles). This means the whole bunch of them will have a mass of about

$$M = 8\,\text{kg} \times 6 \times 10^{23} = 5 \times 10^{24}\,\text{kg}$$

This is about the mass of the Earth or 70 times the mass of the Moon. Sheer lunacy! And with nine lives for each...

And if you think that an Earth made of cats is absurd, continue reading to find out about a Sun made of gerbils (see question 8.5).

What would be the mass of all 10^8 MongaMillions lottery tickets? How many 40-ton trucks would be needed to haul them away?

* * * * * * * * * * * *

HINT: What might be the length of such a ticket? Its width? Remember we estimated its thickness as 2×10^{-4} m.

HINT: What is the area? Convert from square inches to square meters. Four inches is approximately 10 cm or 10^{-1} m, so $(4 \text{ in.})^2$ is approximately 10^2 cm^2 or 10^{-2} m^2.

HINT: Mass = volume × density.

HINT: What might be the density of a ticket compared, say, with that of water? The density of water is 10^3 kg/m^3 (each cubic meter of water has a mass of 10^3 kg or 1 ton).

ANSWER: To estimate the mass of all those tickets, we will need to estimate their volume and their density. The density of an object is expressed in mass per volume. Air has a very low density ($1 \, \text{kg/m}^3$), water has a medium density ($10^3 \, \text{kg/m}^3$ or $1 \, \text{kg/L}$) and lead has a high density ($10^4 \, \text{kg/m}^3$ or $10 \, \text{kg/L}$).

Volume is length times width times thickness. Earlier, we estimated the thickness of each ticket to be $2 \times 10^{-4} \, \text{m}$, so we just need the length and width. A ticket is about 4 inches on a side (more that 1 in. and less than 10 in.). We could choose a slightly different number, but 4 in. $=$ 10 cm, which is a nice round number to calculate squares with. Thus,

$$V = 10 \, \text{cm} \times \frac{1 \, \text{m}}{10^2 \, \text{cm}} \times 10 \, \text{cm} \times \frac{1 \, \text{m}}{10^2 \, \text{cm}} \times 2 \times 10^{-4} \, \text{m}$$
$$= 2 \times 10^{-6} \, \text{m}^3$$

so the volume of 10^8 such tickets is $V = 10^8 \times 2 \times 10^{-6} = 200 \, \text{m}^3$.

What about the mass of this pile of tickets? As we all learned in science class, mass equals volume times density. If we buy a ticket, and don't win, we might be tempted to discard it. Of course, we wouldn't, because it is wrong to litter, no matter how annoyed we might be. But hypothetically, suppose we tossed the ticket into a puddle, would it float or sink? The former, I think, at least until it absorbed some water, and perhaps sank, as some paper products will do after a while. That means that the density of the ticket is reasonably close to the density of water. Since the density of water is $1000 \, \text{kg/m}^3$ or $1 \, \text{ton/m}^3$, the total mass is 200 tons. This would require five 40-ton trucks to haul it away.

Another way of looking at this lottery is that you would need to buy five tractor-trailer trucks full of tickets to make sure that you won!

Tons of trash

How much domestic trash
is collected each year in
the US (in m³ or tons)?

✳ ✳ ✳ ✳ ✳ ✳ ✳ ✳ ✳ ✳ ✳ ✳ ✳ ✳

HINT: How much trash do you throw out each week?

HINT: A kitchen garbage bag is 13 gal (50 L) but can be
compacted.

HINT: Estimate how many homes there are in the US.

ANSWER: Before the era of recycling (and when the children lived at home), we used to empty our 13-gallon trash can in the kitchen about every other day. Now, it's only about once a week, though the recycling bin now gets emptied once a week (with newspapers, boxes and other cardboard containers, bottles, plastic containers, cans, etc.). On second thought, it's easier to lump them all together for this problem. Since considering recycling changes the answer by less than a factor of two for our household, we will ignore it for this problem.

OK, if we empty the trash three or four times a week, that's about 50 gallons of trash for four people. Now a gallon is about 4 liters, and a liter is 10^{-3} m^3, so our 50 gallons per week is 200 liters or 0.2 m^3. In one year (50 weeks), four people produce $50 \times 0.2 = 10$ m^3 of garbage. That is about 300 cubic feet. Yick!

It's worse than that. There are 3×10^8 of us, or about 10^8 households, so we produce 10^9 m^3 of uncompacted garbage.

Now let's try to figure out the mass of that garbage. There are two things to consider here: first, trash is mostly not liquid (ooh, you threw the soup away!), and therefore there's a lot of air space in the trash bag, and second, related to it, the density of the trash is much less than that of water. Let's estimate the density. That full 13-gal (50-L) trash bag probably only weighs between 10 and 20 lb (5 and 10 kg). Thus, its density is between 0.1 and 0.2 kg/L (or 0.1 to 0.2 tons/m^3 or between 10 and 20% that of water).

Let's take an average density of 0.2 tons/m^3. Thus, in one year, my family produced $m = 10$ m$^3 \times 0.2$ tons/m$^3 = 2$ tons of garbage (in US units, that is, um, 2 tons).* Note that, since the average density is

---

* There are lots of different tons: metric, short, long, ... Since they differ by only 10%, we will use them interchangeably.

so low, compacting the trash in garbage trucks should reduce the volume by a factor of about three (more than one and less than five).

But enough about us, let's look at the whole country. With a population of 3×10^8, we produce a total mass and compacted volume of trash of about

$$M = 10^8 \text{ households} \times \frac{2 \text{ tons/year}}{\text{household}}$$

$$= 2 \times 10^8 \text{ tons of trash/year}$$

$$V = 10^8 \text{ households} \times \frac{1}{3} \times \frac{10 \text{ m}^3/\text{year}}{\text{household}}$$

$$= 3 \times 10^8 \text{ m}^3 \text{of trash/year}$$

Now let's compare to reality. According to the US Environmental Protection Agency [9], in 2005 the US generated 245 million (2.45×10^8) tons of municipal solid waste (including recycling).

Now we need to figure out what to do with it all. But that's the subject of the next question.

Mt. Trashmore

3·9

If we put all of that trash (see previous question) in a landfill, how much space will this require? What fraction of the US surface area is this?

* * * * * * * * * * * * * * *

HINT: Is the north–south distance the same as the E-W distance, half, quarter . . . ?

HINT: You can fly from the East coast to the West coast in 6 hours. Alternatively, it is three time zones from NY to LA.

HINT: What is the area of the US?

HINT: How much area do you need for all that trash? How high can you pile it?

HINT: What is the volume of the trash found in the previous problem?

ANSWER: We need to figure out how much area we need for all that garbage and how much area we have available. Let's start with the area we need for the garbage. In the previous question we estimated that Americans generate 3×10^8 m^3 of trash per year. If we pile it 1 m high, we will need 3×10^8 m^2 and if we pile it 10 m high, we will need only 3×10^7 m^2. Here in Virginia Beach we are very proud of our local Everest, Mt. Trashmore, a landscaped former landfill towering 62 ft (20 m) above sea level. This means that we'll need an area of

$$A_{\text{trash}} = \frac{3 \times 10^8 \text{ m}^3/\text{yr}}{20\text{m}} = 10^7 \text{ m}^2/\text{yr}$$

Let's plan ahead and make our landfill large enough for 100 years. In that case, we will need an area of 10^9 m^2. One billion square meters sure sounds like a lot. Let's look more closely. A square kilometer (which is about half of a square mile) is a square 10^3 m on a side, so that $1 \text{ km}^2 = (10^3 \text{ m})^2 = 10^6 \text{ m}^2$. This means that one billion square meters is only(!) one thousand square kilometers ($10^9 \text{ m}^2 = 10^3 \text{ km}^2$). That still sounds like a lot, but it is only the area of Los Angeles* or Virginia Beach and, besides, we have a whole country to dump it in.

What is the US land area? Think of the US as being rectangular in shape. We need to find the east–west (New York to Los Angeles) and north–south (Mexico to Canada) distances. You can fly from NY to LA in six hours. The jet flies at about 500 mph so the distance is about 3000 miles (or 5000 km). Alternatively, we know that there are three time zones from NY to LA so that the distance is $3/24 = 1/8$ of the Earth's circumference at the equator (which we figured out in the golf ball problem). Thus, the east–west distance

---

* Some claim that a landfill would improve LA significantly.

is 4×10^4 km$/8 = 5 \times 10^3$ km. Good, we got the same answer.

Wait a minute, you may say, the US land area does not include the equator! That's no problem (except for ambitious politicians), because the width of three time zones in the US is not that different from that of three time zones at the equator.

We can estimate the "vertical" (i.e., north–south) dimension by just eyeballing the relative proportions of our rectangular US from a map. The north–south span is about one-third of the east–west span or about 1000 miles (1600 km). The land area is then approximately

$$A_{US} = 5 \times 10^3 \, \text{km} \times 1.6 \times 10^3 \, \text{km} = 8 \times 10^6 \, \text{km}^2$$

Thus, the fraction of land area needed for our trash is

$$f = \frac{A_{\text{trash}}}{A_{\text{US}}} = \frac{10^3 \, \text{km}^2}{8 \times 10^6 \, \text{km}^2} = 10^{-4}$$

This means that after we have dumped all of our trash for 100 years in a single huge land fill, we will still have 99.99% of the US land area for everything else we want to do. This is just as Penn and Teller told us [10].

Juggling people

On average, how many people are airborne over the US at any given moment?

* * * * * * * * * * * * * * * * * * *

HINT: The fraction of time people spend flying is equal to the fraction of people flying at any given time.

HINT: Think about the fraction of time you spend flying, e.g., the number of hours or days you fly per year, compared the number of hours or days in a year.

HINT: Don't choose 3:00 AM, choose sometime during the day.

ANSWER: There are two basic ideas here. First, the fraction of time the average person spends flying equals the average fraction of people that are airborne at any instant. This means that if you spend 10% of your time flying, then on average 10% of the population is airborne at any given time.* Note that this only works if there are enough people to average things out.† Second, we can use our own experience to estimate the fraction of time an average person spends in the air, or shopping or sleeping or (you name it). In other words,

$$\frac{\text{number flying now}}{\text{US population}} = \frac{\text{time spent flying}}{1\text{yr}}$$

Back in chapter 1 we estimated that the average American takes between two and four flights per year. The typical flight will take between one and six hours (not counting time spent parking, waiting in lines, consuming the delectable airport comestibles, ...) so we will estimate three flights per year at three hours per flight, or nine hours per year in flight. Now we insert the numbers we know:

$$\frac{\text{number flying now}}{3 \times 10^8 \text{ hr people}} = \frac{9\,\text{hr}}{400\,\text{days} \times 25\,\text{hr/day}}$$

We rearrange this to get

$$\text{number flying now} = 3 \times 10^8 \text{ hr people} \times \frac{9\,\text{hr}}{10^4\,\text{hr}}$$

$$= 3 \times 10^5 \text{ hr people}$$

That means there are about three hundred thousand people airborne over the US at this moment. We hope they all land safely.

---

* This does **not** mean that if you spend 10% of your time flying, that 10% of the average person (that's about one leg) is airborne.
† One other person, or even ten others wouldn't suffice. There have to be enough people so that at every moment some are in the air. This is not a problem for this question.

During the last big California earthquake, two million books fell off the shelves in a university library. How many students would need to be hired to reshelve all of the books in three weeks?

* * * * * * * * * * * * * * *

HINT: How many hours would a student work in a week?

HINT: How many books might a student shelve in an hour?

ANSWER: The books are not just put back on the shelves in random order. They must be placed according to their Library of Congress classification, so care must be taken to identify exactly where the book should reside. We'll assume that the books fell relatively close to where they belong so no one needs to walk across the library. If a book is at my feet, and I know immediately where to put it, that will take between a few seconds and a minute. Thus, we can reshelve between 60 and 600 books per hour. Let's work with an average of 200 per hour (that's one-third of 600 and three times 60).

This means that in three weeks, working eight hours per day and five days per week, one student can reshelve

$$N = \frac{200\,\text{books}}{\text{student-hour}} \times \frac{8\,\text{hr}}{\text{day}} \times \frac{5\,\text{days}}{\text{week}} \times 3\,\text{weeks}$$

$$= 2 \times 10^4 \text{ books/student}$$

Now, we need to reshelve two million books. This means that we need

$$N_{\text{students}} = \frac{2 \times 10^6 \text{ books}}{2 \times 10^4 \text{ books/student}} = 10^2 \text{ students}$$

Thus, 100 students would be needed to reshelve all of those books in three weeks (assuming that they do not stop to read the books as they reshelve them).

Animals and People

Let's continue with some more straightforward questions about animals and people. How big are we, how far do we run, how many porta-potties do we need?

Be warned, the last question is harder.

Chapter 4

How many cells are there in the human body?

* * * * * * * * * * * *

HINT: How might you estimate your volume?

HINT: Volume = mass/density. What is your mass? The density
of water is 1 kg/L or 10³ kg/m³

HINT: Alternatively, estimate your volume as a rectangular box.

HINT: A really big cell is about the smallest object you can see.

ANSWER: Sorry to ask such a personal question, since we hardly know one another, but what's your volume? It's on your driver's licence, along with your surface area ... oops, we were being futuristic. Since you didn't answer the question, let's figure it out. Let's estimate your mass, Mr. Jones, at a nice even 100 kg (or approximately 200 lb; ladies, you can modify this to suit your own figures). Since we can safely assume that you float,* your average density is rather close to that of water or about 1 kg/L or 10^3 kg/m³. Thus, 100 kg of water occupies 100 kg × $(1 \text{ m}^3/10^3 \text{ kg}) = 0.1 \text{ m}^3$. So your volume, Sir, is about 0.1 m³.

We can do this another way. Let's approximate our body as a box of length l, width w, and height h, so our volume is $V = l \times w \times h$. What choices shall we make for these quantities? Length is easy, $l \approx 6$ ft. Now what about w and h? John's cross section is decidedly not rectangular, but since we're not going to say what shape it is,† we'll pick an average width, $w = 1$ ft (remember to average over head, neck, torso, legs, and feet). As for h, the front–back dimension, it will be about, say, 6 in. (he's not barrel-chested). Therefore, his volume is $6 \times 1 \times \frac{1}{2} = 3 \text{ ft}^3$. Now 1 m³ is about $3 \times 3 \times 3 = 27 \text{ ft}^3$, so 3 ft³ is about 0.1 m³.

Now let's figure out the size of a cell using our own eyes. We cannot generally see individual cells with the unaided eye. If you look at a ruler, the lines on the ruler are a fraction of a millimeter (10^{-3} m) wide. I can quite easily see something one-tenth of a millimeter (10^{-4} m) in size. I cannot see cells, so they must be smaller than that. The inventor of the microscope used his first crude microscope (with a magnification between 10 and 100) to see cells. Thus, a typical cell must be 10 to 100 times smaller

---

* That's easy for us to say—we're standing at the side of the pool.

† *Note*: We are dealing with *round* numbers.

than 10^{-4} m or between 10^{-5} and 10^{-6} m (that's 1 to 10 μm [micrometer]) in size.

Here's another approach. We can see cells with an ordinary light microscope.* This means that the cells must be larger than the wavelength of visible light or we couldn't see them. The wavelength of visible light ranges from blue light at about 0.4 μm to red light at about 0.7 μm (400 to 700 nm). Thus, cells must be larger than 1 μm. However, while we can see the major features inside a cell, we cannot see that much detail, so that typical cells must be much smaller than about 100 μm.

A typical human cell with a diameter of 10 μm (10×10^{-6} m $= 1 \times 10^{-5}$ m) will have a volume

$$V_{cell} \approx \text{diameter}^3 = (10^{-5} \text{ m})^3 = 10^{-15} \text{ m}^3$$

Now the number of cells in our body is just the ratio of the volumes:

$$N_{cells} = \frac{V_{body}}{V_{cell}} = \frac{10^{-1} \text{ m}^3}{10^{-15} \text{ m}^3} = 10^{14}$$

This means that you, Sir, and I each have about 100 trillion cells in our bodies. Goodness me! That's about a thousand times as many stars as reside in our galaxy. Start counting . . . and Mr. Lucas? It's time for the first "Cell Wars" trilogy . . .

* A light microscope uses light rather than, say, electrons to "see" with. It may still be rather heavy.

Laboring in vein

4.2

What is the total volume of human blood in the world?

* * * * * * * * * * * * * *

HINT: What is the world population?

HINT: How many pints do you donate at a sitting? What fraction of your blood volume might this be?

HINT: What fraction of your volume is blood?

ANSWER: We can estimate the amount of blood we contain either as a fraction of our total volume or as a multiple of the blood that we can donate. Let's start with our volume. We've already estimated human volume in the previous question; it's about $0.1\,\mathrm{m}^3$ or $100\,\mathrm{L}$. If we were made only of water, our discussion would be more fluid in nature (or we would be all wet). Since the blood has to carry all of the oxygen and other nutrients we need, our body must be more than 1% blood. Similarly, it is almost certainly less than 10% blood. If we take the average, then we each contain about 5% of $100\,\mathrm{L}$, or $5\,\mathrm{L}$ of blood.

Alternatively, we can estimate this from the amount of blood we are allowed to donate. A typical Red Cross donation is about one pint of blood. They would surely not let us donate a dangerous amount of blood, so that this one pint is probably about 10% of our total supply. Now we just need to convert from archaic pints to modern liters. There are two pints in a quart, which is about a liter. Thus, 10 pints is about $5\,\mathrm{L}$, so the two methods give about the same answer. [*Confession*: JA lived in the United Kingdom during the period regarded as dangerous for the incidence of mad cow disease, so he cannot donate blood. But thus far he has shown no symptoms (outside of this book, that is).]

There are about 6×10^9 people in the world. Thus, the total volume of blood is

$$V = \frac{5\,\mathrm{L}}{\text{person}} \times 6 \times 10^9 \text{ people} = 3 \times 10^{10}\,\mathrm{L}$$

There are $1000\,\mathrm{L}$ in a cubic meter, so that $V = 3 \times 10^7\,\mathrm{m}^3$. Now let's see how large a volume this is.

New York City's Central Park is about $2\,\mathrm{km}^2$ or about $2 \times 10^6\,\mathrm{m}^2$. Thus, this would cover Central Park to a depth

$$d = \frac{\text{volume}}{\text{area}} = \frac{3 \times 10^7\,\mathrm{m}^3}{2 \times 10^6\,\mathrm{m}^2} = 15\,\mathrm{m}$$

This is 50 ft or the height of a 5-story building. Given the number of murders occurring in TV series like *CSI New York* and *Law and Order*, one might be forgiven for thinking that most of this blood is already there!

If we want to get biblical, we can compare it to the volume of blood shed at the battle of Armageddon as mentioned in the book of Revelation: "They were trampled in the winepress outside the city, and blood flowed out of the press, rising as high as the horses' bridles for a distance of 1,600 stadia" Rev. 14:20 (NIV). A horse's bridle is about 2 m high. At almost 200 m per stadion, 1600 stadia is 300 km. Now we just need the width. That much liquid will spread out a lot, especially when flowing 300 km. Let's use a width of 3 km. Thus, the volume of blood predicted to flow at Armageddon is

$$V_{\text{Armageddon}} = 2 \text{ m} \times 3 \times 10^5 \text{ m} \times 3 \times 10^3 \text{ m}$$
$$= 2 \times 10^9 \text{ m}^3$$

That is about 15 times more blood than humans currently have. We guess we just need more people.

Unzipping your skin

4·3

What is the surface area of a typical human?
(Include only the skin, not the surface area of the
digestive tract.)

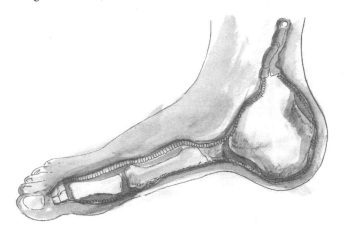

HINT: Think of yourself as a flat sheet and estimate your length
and width. Don't forget the area of your back.

HINT: Alternatively, imagine you are a cylinder and estimate your
height and radius. The surface area of a cylinder (ignoring top
and bottom) is $2\pi r$ times radius times height. If you're interested,
its volume is $\pi r^2 h$ (its volume is still $\pi r^2 h$ even if you're not
interested).

ANSWER: If you could unzip your skin (like the rhinocerous in Kipling's *Just-So* story), how large an area would it cover? The area of a sheet on a double bed? Your backyard? This is actually an important question since it determines how much fabric you need to make a suit of clothing, how much sunlight hits your body, how much force is exerted on you when scuba diving, how much of certain medicines you should take, and how much you can perspire.

You could measure most of your area by taking a shirt and a pair of pants, disassembling them, and measuring their total area, but we are not trying to be that precise. There are at least two reasonable approximations, humans as cylinders and humans as flat sheets.

We are certainly not cylindrical critters of height h and radius r, but if we were, then our surface area would just be $A = 2\pi rh$. We can account for the fact that we are not simple cylinders by multiplying the final answer by 1.5 or 2. Let's estimate r and h. Height is easy, ours is 2 m (well, let's just say that it rounds up to 2 m [6 ft 7 in.]). Radius is less than 1 m (3 ft) and more than 0.1 m (4 in.) so we'll estimate 0.5 m. This gives a surface area of

$$A = 2\pi rh = 3 \times 0.5\,\text{m} \times 2\,\text{m} = 6\,\text{m}^2$$

We can check to see if this is consistent with our previous volume estimate of $0.1\,\text{m}^3$. The volume of a cylinder is the area of its base (πr^2) times its height, or $V = \pi r^2 h$. This gives $V = 3 \times (0.5\,\text{m})^2 \times 2\,\text{m} = 1.5\,\text{m}^3$.

Oops, that is way too much. Let's decrease the radius to 0.2 m (8 in.). That will give us a volume of $0.24\,\text{m}^3$, still a bit too big, and an area $A = 2.4\,\text{m}^2$. Since we're not really cylinders, let's round the answer up to $3\,\text{m}^2$.

That's about the area of a medium-sized towel (not including the fibers!). Of course, we have lots of curves and corners, so this is just a rough estimate, even if we shaved first.

Alternatively, suppose we were flat like a sheet; front and back would each be about 2 m (6 ft) high by 0.5 m (1.5 ft) wide, so that's a total of 2 m^2, a bit smaller than the first estimate, and much simpler to obtain!

Body surface area (BSA) is important in medicine for calculating the appropriate dose of some medications. You can calculate your own BSA at many websites [11]. I calculated mine to be 2.06 m^2, so our estimates are pretty darn good (for purposes of this book, but certainly not for medical purposes).

At ten tons of air pressure on every square meter, there are 20 tons of air pushing on you. If you scuba dive down to 120 ft (40 m), there are another 40 tons of water pressure on every square meter. That adds up to another 80 tons of water pressing on you. It's amazing that the human body can withstand those forces!

Hair today, gone tomorrow

What is the total length
of all the hair on an
average woman's
head?

* * * * * * * * * * * * *

HINT: It is easier to do the arithmetic if you convert to metric by
remembering that 4 in. = 10 cm.

HINT: How long is each strand of hair?

HINT: What is the average separation between hairs?

HINT: What is the area of her scalp?

ANSWER: We will answer this question in three stages. First we will estimate the area of the scalp, then the number of hairs per square centimeter, and then the length of a typical strand of hair.

My hand spans a distance of about 8 inches or 20 cm. My head is about one hand span in diameter. Assuming my head to be a sphere (and my scalp to be a hemisphere), the area of my scalp is

$$A_{scalp} = \frac{1}{2} 4\pi r^2 = 6(10 \, cm)^2 = 600 \, cm^2$$

(Alternatively, if you treated your scalp as a square 20 cm on a side, you would have gotten 400 cm². Close enough!)

Now we need the number of hairs per area (per square centimeter). We can count the number of hairs along one centimeter and square it. There are 1–2 hairs per millimeter or 10–20 per centimeter. This gives 100–400 hairs per square centimeter. We'll use 200. This means that the total number of hairs on a person's head is about

$$N = 6 \times 10^2 \, cm^2 \times 2 \times 10^2 \, hairs/cm^2$$
$$= 10^5 \, hairs$$

Blond hair is typically finer and more closely spaced than black hair so your answers may vary. However, this unnecessary precision amounts to bifurcating rabbits.*

All we need now is the length of a typical hair. Women's hair ranges from 1 cm (0.5 in.) to 100 cm (1 m or 3 ft). The geometrical mean of 1 and 100 cm is 10 cm (4 in.) or about shoulder length. This means

---

* Splitting hares.

that the total length of all the hairs on a woman's head is about

$$L = 10^5 \text{ hairs} \times 10 \text{ cm/hair} = 10^6 \text{ cm} = 10^4 \text{ m} = 10 \text{ km}$$

This is 10 kilometers or about 6 miles. Short hair would be about 1 km and really long hair would be about 100 km.

It's enough to make your hair stand on end.

Hot dawg!

4.5

How long a hot dog (or sausage or wurst or . . .)
can be made from a typical cow?

HINT: How long would that hot dog have to be so that its
volume is the same as the cow's?

HINT: What is the thickness of a hot dog?

HINT: What is the volume of a human?

HINT: How much larger is a cow than a person, 10 times?
100 times? 1000 times?

ANSWER: We first need to figure out the volume of a cow. A cow is about ten times the weight of a person (about one ton or the weight of a small car). It is definitely less than 100 times the weight of a person (10 tons is the weight of a large truck). Since the densities are about the same, this means that it has ten times the volume of a person or $V_{\text{cow}} = 10 \times 0.1 \, \text{m}^3 = 1 \, \text{m}^3$.

Now we just need to estimate the volume of a sausage or hot dog. We'll assume square hot dogs so we don't to bother with pesky factors of π. The thickness, t, of a typical sausage or "dog" is 1 in. (2 cm or 0.02 m). Thus, for a hot dog of length L, it will have a volume of

$$V_{\text{hotdog}} = L \times t^2$$

and since it must have the same volume as the cow it is made from, its length will be

$$L = \frac{V_{\text{cow}}}{t^2} = \frac{1 \, \text{m}^3}{(2 \times 10^{-2} \, \text{m})^2}$$
$$= 2 \times 10^3 \, \text{m}$$

This is 2000 m or 2 km (over 1 mile)! Wow!

Note that this assumes that every little bit of the cow is ground up into hot dog. We really, really hope that this is a bad assumption.

A hot dog made from a human being would be ten times shorter or about 200 m. That is twice the length of a football field!

How far does a soccer or field hockey player travel during the course of a 90-minute game?

* * * * * * * * * * * * * *

HINT: Distance covered = average speed × time spent moving.

HINT: Alternatively, how fast does she run and how much of the game does she spend running?

HINT: How many times does she run up and down the field?

ANSWER: If you've ever watched that most wonderful of sporting events, the World Cup,* you'll have noticed that players (except the goal keepers) spend most of their time running or walking, only occasionally standing still (such as when some reckless and violent individual playing against England commits a terrible foul against an innocent player who would otherwise have scored a magnificent and game-winning goal). The players rarely stand still; they sometimes walk and sometimes run quite fast.

Let's estimate walking and running speeds. The world record for running the mile is about 4 minutes. That would be about 15 miles per hour, which, at $1 \text{ m/s} \approx 2 \text{ mph}$, is about 7 m/s. Alternatively, the world record for the 100-m dash is about 10 s, or about 10 m/s. Walking speeds are closer to 2–4 mph or 1–2 m/s. If we assume that half the time is spent running at full speed and the other half is spent walking, then we'll have an average speed of $(7 + 1)/2 = 4 \text{ m/s}$ or 8 mph. (The correct answer is surely between 2 and 8 m/s so our answer must be within a factor of two!)

During a standard 90-minute (1.5-hour) game, the players travel about 8 mph times $1.5 \text{ hr} = 12 \text{ mi}$ (or almost 20 km). Quite a workout!

* *Disclaimer*: This is not the opinion of all the authors and certainly not of the editors.

Ewww ... gross!

4·7

How many people in the world are picking their nose right now?

✳ ✳ ✳ ✳ ✳ ✳ ✳ ✳ ✳ ✳ ✳ ✳ ✳ ✳ ✳

HINT: There are 6×10^9 people in the world.

HINT: The fraction of people picking their nose equals the fraction of time each them spends in that indelicate activity.

HINT: There are about 10^3 minutes in a day; how many of them does a typical person spend picking his or her nose?

ANSWER: As with the flighty Americans problem, the fraction of the time you or I spend in some activity (except for scuba diving: none!) is equal to the fraction of people doing it right now.*

We won't ask how long you spend picking your nose each day, but how long do you estimate that your friends spend in this activity? Ten seconds is probably far too short, and 1000 seconds or 15 minutes is too long, so 100 seconds (about 2 minutes) seems a reasonable compromise. We'll further assume that our children are correct when they claim that "Mom, everybody does it!"†

Don't tell our children, but there is some evidence that ingestion of nasal mucous helps train the immune system to recognize harmful bacteria and viruses.

Anyway, we'll estimate that all six billion of us spend an average of 2 minutes per day picking our nose (neglecting people with more than one nose). There are about $25 \times 60 = 1500$ minutes in a day. Thus, we have

$$\frac{\text{number picking now}}{6 \times 10^9 \text{ people}} = \frac{2 \text{ min}}{1500 \text{ min}}$$

or

$$N_{\text{pick}} = 6 \times 10^9 \text{ people} \times \frac{2 \text{ min}}{1500 \text{ min}}$$

$$= 10^7 \text{ people}$$

Thus, ten million people are picking their nose at this very moment. Hmmm . . . so this is why politicians use up gallons of hand sanitizer. They shake way too many hands.

---

* Assuming, of course, that the authors and readers of this book are typical people. Hah!

† Videotape of an audience of doctors at a medical lecture indicated that our children's claim is quite correct.

Going potty

How much space would
a million people need
at a political rally?

How many porta-potties?

4.8

* * * * * * * * * * * * * * * * *

HINT: How much elbow room do you have at a political rally?
How many people are crammed into a typical square meter?

HINT: The fraction of people in the porta-potty equals the
fraction of time each spends in that indelicate activity.

HINT: What fraction of the time do you spend in the loo?

ANSWER: In a political rally, people are packed reasonably tightly together. There is still space to move around, so we're not packed as tightly as in the very first question. Let's estimate that there's about two feet between people. Since each person is about one foot in size, each person occupies a space that is about 3 ft by 3 ft or $1\,m^2$. That means that 10^6 people occupy $10^6\,m^2$ or $1\,km^2$. That is about the size of the National Mall in Washington, DC or Central Park in New York. Now that you have all those people there, you need to provide facilities for inputs and outputs, that is to say refreshments and porta-potties.

Again, as with the flighty Americans or the nose picking, the fraction of the time I spend in some activity (except for bungee-jumping: none!) is equal to the fraction of people doing it right now. So: how long do you spend in the bathroom during the daytime? Ignore time spent grooming or powdering your nose. One minute is probably too short and 100 minutes is far too long ("Are you going to be in there ALL day?"). So let's settle on 10 minutes, shall we? Also, let's just take a 15-hr day because most people don't go to political rallies while they are sleeping. Let's convert 15 hours to $15 \times 60 = 10^3$ minutes (so it has the same units as our bathroom interval). Thus, the fraction of daytime spent in the bathroom (and hence the fraction of people in the bathroom at any instant) is 10 minutes out of 10^3 minutes or 1%.

This means that we need one potty for every 100 people, or a total of 10^4 potties for 10^6 people. Wow, that's a huge logistical effort! Note that if there is one potty for every 100 people, given that I spend 1/100 of my time there, then all the potties will be 100% occupied during the entire rally. Since the demand for potties is generally not uniform, there will be serious queuing problems (that is British for "really long lines"). So a few more might be advantageous...

Now let's compare with reality. The Virginia Children's Festival attracts about 3000 people annually to Norfolk's Town Point Park. They have about forty or fifty porta-potties there. The number of potties required, according to our estimate, is $3000/100 = 30$, so they have a few extra to reduce the waiting times.

Let's get one thing straight!

4·9

How long is all the DNA
in your body?
How long is the DNA
of all humanity?

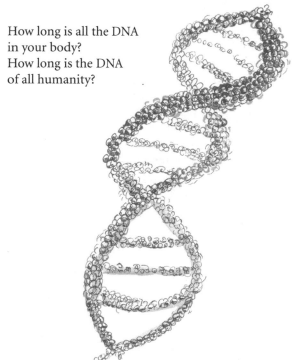

❋ ❋ ❋ ❋ ❋ ❋ ❋ ❋ ❋ ❋ ❋ ❋ ❋ ❋ ❋ ❋

HINT: How much DNA is in one cell?

HINT: DNA is composed of long strings of "base pairs," each
composed of about 1000 atoms.

HINT: One atom is about 10^{-10} m in size.

HINT: Each base pair is approximately a cube of side length
10^{-9} m.

HINT: The nucleus of a cell is filled with DNA and is about 1/10
the size of the cell.

HINT: We just estimated the number of cells several questions
ago.

83

ANSWER: This one is a little complicated. We need to calculate the size of the building blocks of DNA and the total volume of DNA in the cell. Then we can use that to figure out how long the total DNA is.

The nucleus of a cell is filled primarily with long strands of DNA called chromosomes.* Each chromosome is composed of long strings of base pairs (the familiar letters $ATGC$ from a long forgotten biology course). Each base pair is a very complicated molecule itself. This means that it must contain a lot of atoms, certainly more than 100 and less than 10^4, so we'll estimate 10^3 atoms per base pair. We will treat each base pair as a cube, 10 atoms on a side. Since all atoms are about 10^{-10} m, our base pairs are 10^{-9} m in length. The volume of a base pair is then the length cubed:

$$V_{bp} = (10^{-9} \text{ m})^3 = 10^{-27} \text{ m}^3$$

We estimated earlier that a cell is about 10^{-5} m in size. The nucleus is about $1/10$ of that, or 10^{-6} m. Thus, the volume of the nucleus is

$$V_n = (10^{-6} \text{ m})^3 = 10^{-18} \text{ m}^3$$

This means that each nucleus can contain a number of base pairs

$$N_{bp} = \frac{V_n}{V_{bp}} = \frac{10^{-18} \text{ m}^3}{10^{-27} \text{ m}^3} = 10^9$$

That is one billion base pairs. That's a LOT of information.

Now let's straighten out all that DNA. There are 10^9 base pairs at 10^{-9} m each so the total length of DNA in the cell is about 1 m. According to biology textbooks, the length is between 1 and 3 m [12, 13].

---

* This is a physicist's view of a cell and is accurate to within a factor of two or three. If you want more precision, call a friendly biologist.

Now we estimated previously that there are 10^{14} cells in our bodies. At 1 m of DNA each, that stretches out to 10^{14} m. That is 1000 times the distance from the Earth to the Sun, ten times the distance from here to Pluto, or about 1% of a light year.

If we stretched out the DNA of everyone on the planet, the combined DNA strand would extend about 10^8 light years—about 50 times farther than the next galaxy! But who would be around to appreciate the comparison?

Transportation

Americans have a love–hate relationship with their cars. We love the freedom and privacy they provide but we hate the traffic and pollution they cause and the expensive fuel they need. In this chapter we look at how far we drive, how much it costs, and what some of the alternatives are.

Chapter 5

✳ ✳ ✳ ✳ ✳ ✳ ✳ ✳ ✳ ✳ ✳ ✳ ✳ ✳ ✳ ✳

How many total miles (or kilometers) do all
Americans drive in one year? How does this compare
to the circumference of the Earth (2.5×10^4 mi),
the distance to the Moon (2.4×10^5 mi), the distance
to the Sun (9×10^7 mi), or the distance to
Saturn (10^9 mi)?

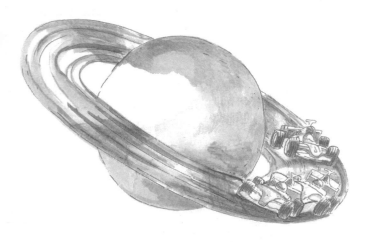

HINT: How far does one person drive in a year?

HINT: How many miles per year do new car warranties cover?

HINT: There are 3×10^8 Americans.

ANSWER: To answer this question we need (1) how many miles each American drives and (2) the total number of Americans who drive. There are many ways to estimate our average mileage. You could look it up, but that would violate the spirit of this book, and besides, it would involve getting up and walking over to the computer. You could use your typical yearly mileage (which would probably be within a factor of two of the national average). Or you could look at new car guarantees. Manufacturers advertise 3-year/36,000-mile or 10-year/120,000-mile warranties. This indicates that each car is driven about 12 thousand miles per year.

Now we need the number of cars. There are 3×10^8 Americans. Almost all of us drive. There is less than one car per person and more than one car per four people so we will estimate one car for every two people.* Thus, the total miles driven by all Americans is about

$$d = \frac{10^4 \text{ mi}}{\text{car}} \times 3 \times 10^8 \text{ people} \times \frac{1 \text{ car}}{2 \text{ people}} = 2 \times 10^{12} \text{ mi}$$

That is 2 trillion miles or 3 trillion kilometers. Wow! That distance would take us around the Earth 10^8 times, to the Moon and back 4 million (4×10^6) times, or to Pluto and back 2000 times. It takes the Earth 3000 years to travel that distance in its orbit around the Sun. At a speed of 3×10^8 m/s, it takes light one-third of a year to travel that far.†

Who knew it was that far to the corner store!

---

* In 2000, there were an average of 2.6 people and 1.3 cars per American household. This is fortunate because only the 0.6 person can drive the 0.3 car [14].

† This means this means that Americans travel at a total speed of only one-third that of light.

Drowning in gasoline

5.2

What volume of gasoline does a typical automobile (car, SUV, or pickup) use during its lifetime? Note that this question asks about the lifetime of the vehicle, not the time that *you* own it. Compare the weight of the fuel to the weight of the car.

✳ ✳ ✳ ✳ ✳ ✳ ✳ ✳ ✳ ✳ ✳ ✳ ✳ ✳ ✳ ✳

HINT: How many miles does a typical car drive in its lifetime? 10^1? 10^2? 10^3? 10^4? 10^5? 10^6?

HINT: How many miles does it get per gallon?

HINT: There are four liters per gallon and 10^3 liters per cubic meter.

HINT: The density of water is one ton per cubic meter.

ANSWER: First, let's estimate how many total miles a vehicle travels in its lifetime. Your vehicle's odometer typically has six digits. This indicates that manufacturers expect your vehicle to last at least 100,000 (10^5) miles. However, most vehicles do not reach much past 200,000 (2×10^5) miles.* Let's start with 10^5 miles because it is a nice round number. (Cars made in the 1950s and 1960s did not last as long and frequently had only five-digit odometers.)

Now let's estimate the gas mileage. There are several ways to do this. You can have an obsessive attention to detail and calculate your mileage every time you fill your gas tank. You can read the new car ads or *Consumer Reports* and notice the mileage. Or you can estimate it from how far you travel between fill-ups and how much gas you put in. Most cars travel about 300 miles between fill-ups. If you have to put in 10 gallons of gasoline, then you get

$$\frac{300\,\text{mi}}{10\,\text{gal}} = 30\,\text{mpg (miles per gallon)}$$

If you have to put in 30 gallons, then you get only $(300\,\text{mi})/(30\,\text{gal}) = 10\,\text{mpg}$. Since almost all cars get between 10 and 40 mpg, we'll use 20 mpg for our estimates.[†]

Now we can calculate the answer. The total number of gallons used is equal to the total miles driven divided by the miles per gallon:

$$V_{\text{gas}} = \frac{10^5\,\text{mi}}{20\,\text{mpg}} = 5 \times 10^3\,\text{gal}$$

---

* (Larry here:) I tried to keep my previous car until it reached the moon, but had to get rid of it at only 225 thousand miles, not the 240 thousand I was aiming for.

† This way, we'll be within a factor of two of the right answer for almost any vehicle.

Now we need to convert to metric, since converting to cubic inches or feet or yards would be too complicated. One gallon equals four quarts, which equals approximately four liters. Thus, 5×10^3 gal $\approx 2 \times 10^4$ L. Since there are 1000 L in a cubic meter, this gives a volume of gasoline equal to 20 m^3.

Now let's see how big that is. A 20-m^3 volume could be 1 m deep by 4 m wide by 5 m long. At 3 ft per meter (actually 3.3), that's 3 ft deep by 12 ft wide by 15 ft long. That is much larger than your car. It's enough to fill a small above-ground swimming pool!

At a density of 1 ton/m^3, your car will burn about 20 tons of fuel in its lifetime. That is ten times the weight of the car itself! Even if you drive a Prius or other car that gets 50 mpg, you will still burn 8 tons of fuel.

Maybe your car lasts longer than 10^5 miles. Maybe your car is a 10-mpg gas guzzler or a 40-mpg gas sipper. Your answer will still be well within a factor of ten. The fuel your car burns will greatly outweigh the car itself.

How much total extra time would Americans spend driving each year if we lowered the highway speed limit from 65 to 55 mph? (Note that we assume that there is some relationship between posted limits and actual speeds on highways.)

Give your answer in lifetimes.

* * * * * * * * * * * * * *

HINT: How many total miles do we drive? (See previous question.)

HINT: What fraction of those miles do we drive on the highway and what fraction of those do we drive on 65-mph highway?

HINT: Use 60 and 70 mph instead of 55 and 65 to make the arithmetic easier.

ANSWER: We need to estimate the time we spend driving on 65-mph highways. To do this, we need to estimate the *distance* we drive on 65-mph highways.

From the answer to an earlier question, we know we drive 2×10^{12} (that's two trillion) miles per year. About half of that is spent driving on the highway. Urban highways do not have 65-mph limits (as if you could ever drive that fast on the Santa Monica Freeway or New York's West Side Highway). We'll estimate that only about half of highway driving is done on 65-mph roads. Thus, we drive

$$d_{65\,\text{mph}} = 2 \times 10^{12} \text{ mi} \times \frac{1}{2} \times \frac{1}{2} = 5 \times 10^{11} \text{ mi}$$

on 65-mph roads.

Now we need to figure out (1) how much time it would take to drive at a 65-mph speed limit, (2) how much time it would take to drive at a 55-mph speed limit, and (3) the difference. Instead of using 65 and 55 mph, we will use 70 and 60 mph. We do this for two reasons: (1) No one knows what the real (i.e., the enforced) speed limit is so most traffic exceeds the limit at least somewhat, and (2) it is much easier to do arithmetic with round numbers.

At 70 mph, driving takes a total time of

$$t_{65\,\text{mph}} = \frac{5 \times 10^{11} \text{ mi}}{70 \text{ mph}} = 7 \times 10^9 \text{ hr}$$

If the speed limit is lowered to 55 mph and people drive at 60 mph, then

$$t_{55\,\text{mph}} = \frac{5 \times 10^{11} \text{ mi}}{60 \text{ mph}} = 8 \times 10^9 \text{ hr}$$

Thus, we will spend an extra time

$$t_{\text{extra}} = t_{55\,\text{mph}} - t_{65\,\text{mph}} = 8 \times 10^9 \text{ hr} - 7 \times 10^9 \text{ hr} = 10^9 \text{ hr}$$

That is an extra one billion hours behind the wheel. Yikes!

There are only $24 \times 365 \approx 10^4$ hours in a year. Thus, the extra 10^9 hours is an extra 10^5 *years*.

At 100 years in a lifetime, that is an extra 1000 *lifetimes*.

Holy cow! That is a lot more lifetimes than we lose to sharks or lightning or secondhand smoke or arsenic in drinking water or . . . .

Of course, there is a huge difference between one person losing her entire life or millions of people losing a few hours each. (After all, 10^9 hr divided among 3×10^8 Americans is only 3 hr each.)

Note that we have not asked how many lives would be saved by lowering the speed limit. Since traffic deaths depend on actual speed driven, speed differences among drivers, and many other factors, it is very hard to predict the change in fatalities. There was no obvious change in the high fatality rate when the nationwide 55-mph limit was repealed and different regression studies reach different conclusions. Some claim that raising the limit saved lives.

What are the relative costs of fuel (per kilometer or per mile) of New York City bicycle rickshaws (human-pedaled taxis) and of automobiles?

* * * * * * * * * * * * * * * * *

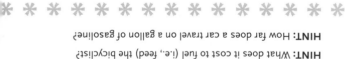

HINT: How far does a bicycle rickshaw travel in one day?

HINT: What does it cost to fuel (i.e., feed) the bicyclist?

HINT: How far does a car travel on a gallon of gasoline?

ANSWER: Let's start with automobiles. If your car gets 20 miles per gallon and gas costs $3.00 per gallon, then it will cost you

$$\frac{\$3.00/\text{gal}}{20\,\text{mi/gallon}} = \$0.15/\text{mi} = \$0.10/\text{km}$$

Cars are cheap to run!*

Now let's figure out how much the human costs per kilometer. To do that, we need to estimate the costs and the kilometers (or miles). Bicycling speed with passengers will be between 5 (very slow) and 10 (fast) mph. Thus, during an 8-hour day, the rickshaw can travel between 40 and 80 miles. Since it is not always in motion, we'll use the lower estimate of 40 miles (or 60 km). We can also consider the performance of experienced bicyclists. They can travel 100 miles in a day. Pulling passengers and dealing with NYC traffic would certainly slow them down to 50 miles (or less).

Working that hard will burn a lot of calories.[†] We can estimate the costs of food several ways, from per diem rates for expense accounts, from fraction of annual income, from the costs of fast food, etc. Let's see what we get.

Expense account per diems are about $40. This is almost certainly too high, since it applies primarily to business folks who expect to eat three meals a day at good to excellent restaurants. Western European or American per capita income is $30,000–40,000 per year or about $100 per day. We spend more than 5% and a lot less than half of our income on food, so we'll estimate we spend 10–20% of our income or $10–20 per day on food. Buying three meals a day in fast-food

---

* Since the difference between miles and kilometers is only a factor of 1.6, it really does not matter which one you use.

† We'll investigate food as an energy source in Chapter 5.

restaurants would also cost about $10–20 per day.* Thus, the fuel costs for the rickshaw pedaler will be

$$\frac{\$15/day}{60\,km/day} = \$0.25/km$$

Thus, the rickshaw costs only 2–3 times more than the car for fuel. The car uses more energy per mile, but gasoline is a much cheaper energy source than food.

Note also that the fuel costs for the autmobile depend strictly on distance traveled (i.e., if you double the distance, you double the fuel cost) but that the fuel costs for the rickshaw bicyclist include the 2000 Calories per day just for basic metabolism.

---

* Since these are New York City rickshaw drivers, we will assume Western diet and food costs.

Horse exhaust

How much waste is generated (per kilometer) by horse-drawn carriages and by automobiles? Give your answer in kg/km.

HINT: Horse exhaust is either solid or liquid and car exhaust is gaseous.

HINT: Assume that all the input (food and fuel) becomes output (waste).

HINT: How far does a horse-drawn carriage travel in one day? Horses are not that much faster than people.

HINT: How much does the horse eat and drink? A horse is about 10 times as massive as a person.

HINT: How far does a car travel on a gallon of gasoline?

ANSWER: We'll need to first estimate how far a horse can pull a carriage in one day and then estimate how much food and water it consumes in the process. Note that we are assuming that the horse does not gain or lose weight and therefore it converts its food and drink into an equal mass of horse exhaust.

How far can an average horse pull an average carriage in one day? Since we are making the horse work for about eight hours, it is certainly not going to gallop, canter, or trot for most of it. Let's start with the average walking speed of a horse. A horse will certainly walk faster than a person (3–4 mph or about 5–6 kph or about 2 m/s) but not that much faster. This gives us a range of about 5–10 mph. In 8 hours, it can travel 40–80 miles, so we'll choose 60 miles (or 100 km).*

How can we figure out the food consumption? We can try to directly estimate the food consumed or we can start with a human and scale up. A horse will certainly eat more than 1 quart (or liter) of grain and less than 100 quarts. We'll take the geometrical mean and estimate 10 L. That has a mass of about 10 kg (20 lb)† and will therefore produce 10 kg of waste. Similarly, the horse will drink more than 1 L and less than 100 L so we will estimate that it drinks about 10 L (also 10 kg).

Now let's start with a human and scale up. We eat 2–3 lb (1–1.5 kg) of food and drink 1–2 L (or quarts) of liquid per day. A horse is about ten times our size (in volume or mass) and thus probably consumes about ten times more than we do. This gives about the same consumption as the previous estimate.

---

* That factor of 1.6 is almost irrelevant. Our estimates are just not that exact.

† The density of anything organic is reasonably close to water. The density of water is 1 g/cm^3 or 1 kg/L or 1 ton/m^3.

Thus, the horse will produce about 10 kg each of liquid and solid waste in the course of traveling 60 miles. The total waste will be about 0.3 kg/mi.

The automobile burns about one gallon every 20 miles. A gallon is about 4 L and thus has a mass of about 4 kg. Thus, the car produces 4 kg of waste in 20 miles or 0.2 kg/mi.

A car produces about the same amount of waste per mile as a horse.* The problem is that while the car's exhaust blows away into the air (and is almost entirely carbon dioxide and water), the horse's exhaust sticks around (literally). Imagine New York City if it had millions of horses instead of millions of cars. Better yet, don't.

* Both produce plant food! The car produces carbon dioxide and the horse produces fertilizer.

Tire tracks

How far does a car travel
before a one-molecule
layer of rubber is
worn off the tires?

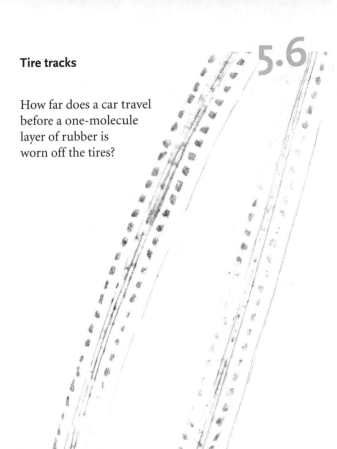

5.6

* * * * * * * * * * * * * * * *

HINT: What is the lifetime of a tire (in miles)?

HINT: How thick is the tread of a new tire? How much of this is
worn away during a tire's lifetime?

HINT: A rubber molecule is only a few tenths of a nanometer
(i.e., a few ×10^{-10} m) in size.

ANSWER: First we need to figure the lifetime of tires in miles. As usual, there are a few ways to do this. You can estimate the lifetime of a tire in years and assume the usual 12,000 miles per year. Tires definitely last more than 1 year and less than 10, so estimates of between 3 and 5 years are reasonable. Alternatively, you can read the tire ads, which advertise the tire lifetimes, or remember the lifetime of the last set of tires you bought. Tires typically last 30–60 thousand miles. They typically have between 1/4 and 1/2 in. (i.e., about 1 cm) of tread.

Thus, 1 cm of tread is worn off in about 4×10^4 mi. We want to know how long it takes to wear off a thickness of one molecule or 5×10^{-10} m of tread. That distance is

$$
\begin{aligned}
d &= \frac{4 \times 10^4 \text{ mi}}{1 \text{ cm}} \times \frac{100 \text{ cm}}{1 \text{ m}} \times 5 \times 10^{-10} \text{ m} \\
&= 20 \times 10^{-4} \text{ mi} \\
&= 2 \times 10^{-3} \text{ mi}
\end{aligned}
$$

Now we need to make sense of this result. 10^{-3} mi is hard to figure out, but 10^{-3} km is just 1 m. Since a mile is only a little bigger than a km, we have

$$
d = 2 \times 10^{-3} \text{ mi} = 3 \times 10^{-3} \text{ km} = 3 \text{ m}
$$

Three meters is about 10 feet. That is only one or two complete rotations of the tire.

Thus, you wear off a one-molecule thickness of rubber with every rotation of your tire.

Working for the car

Your car allows you to travel many miles
in just a few hours. However, in addition
to the hours you spend driving, you
have to spend more hours not driving,
hours you spend earning money to pay
for your car (eg: depreciation,
insurance, fuel). This extra time
reduces your average car travel
speed. For example, if you drive
60 miles in one hour and then
spend one more hour earning
enough money to pay for the
driving, then your average
speed is not 60 mph,
but 30 mph.

If you add all the time you
spend working in order to
earn the money to pay for
your car to all the time that
you spend driving your car,
what is your average car
travel speed?

HINT: What is the total cost of the car over its lifespan?
Include purchase price, insurance, repairs, gas, maintenance,
parking, · · · · . Alternatively, you can use the IRS mileage
reimbursement rate, which should include all of these costs.

HINT: How far do you drive a car during its lifespan?

HINT: How many hours does it take you to drive all those all those miles?

HINT: What is the average American wage? (Note that there are
2000 working hours in a year [50 weeks/year × 40 hours/week] so
that a salary of $40,000 per year is equivalent to $20 per hour.)

ANSWER: We need to figure out how much time we spend working to pay for the car and how much time we spend driving the car. Let's put the pedal to the metal and take it from the top.

To figure out the working time, we need to first estimate the life-cycle cost of the car. There are at least two ways to estimate the cost of a car. We can take the IRS mileage reimbursement rate of $0.445 per mile in 2006. That's the easy way. (As usual, any value between $0.30 and $0.60 is quite reasonable.) Alternatively, we can try to figure out the life-cycle cost of a car. Here we go.

Let's buy an average new car. No Hummers, no Jaguars, no Minis. Then let's keep the car until it gets old or about ten years. We will have to pay for many things in those ten years: the car, insurance, gas, repairs, fuzzy dice, and parking.

The best selling vehicle in America in 2006 was the Ford F-150 pickup truck and the best selling car was the Toyota Camry. Both vehicles cost between $20,000 and $25,000 (depending on options and ability to bargain).

Insurance will cost about $1000 per year or $10,000 for 10 years. It will be a lot more for an 18-year-old male driving a Corvette with three speeding tickets and a DUI in the Bronx. It will probably be much less for a 45-year-old woman with a perfect driving record in Salem, Oregon.

Repairs will probably average about $1000 per year. Repairs will cost much less in the first few years and much more in the last few years.

In ten years, we will drive about 120,000 miles (or about fives times around the Earth). At the Camry's 20 miles per gallon, we will burn $\frac{120,000}{20} = 6000$ gallons of gasoline. At $3 per gallon, that will cost $18,000. The difference in mileage between the Camry and the

F-150 will amount to "only" a few thousand dollars of gas and is thus too small for this book to worry about.

If you live in Manhattan (New York, not Kansas), you can add $3000 per year for parking. That's your problem. Most of us can leave that out.

Thus, the total costs of driving 1.2×10^5 mi in 10 years is (to one significant figure):

Category	Cost ($)
Vehicle cost	20,000
Insurance	10,000
Repairs	10,000
Gasoline	18,000
Fuzzy dice	5
Total	60,000

If we use the IRS mileage number of about $0.50 per mile, we also get a total cost of $60,000 for 120,000 miles.

Note that the total (life-cycle) cost of the car is three times the purchase price.

Now we need to figure out (1) how much time we spend driving the 120,000 mi and (2) how much time we spend earning the $60,000.

Most people typically drive on a mix of highways (55–65 mph) and local streets (25–45 mph plus traffic lights and stop signs). If we average highway and local (or Montana and New York City), we'll get a speed of 30–40 mph. This means that we will spend a total of

$$t_{drive} = \frac{1.2 \times 10^5 \text{ mi}}{40 \text{ mph}} = 3000 \text{ hr}$$

behind the wheel to travel 120,000 miles. If you average 30 mph, you will spend 4000 hours.

Per capita American income is $40,000 per year. Of course, we don't spend all of our time working (it only seems that way). In one year, we work

$$t_{work} = \frac{40\,hr}{week} \times \frac{50\,weeks}{year} = \frac{2000\,hr}{year}$$

Since it takes 2000 hours to earn $40,000, we earn an average of $20 per hour.* At $20 per hour, it will take

$$t_{earn} = \frac{\$60,000}{\$20/hr} = 3000\,hr$$

Wow! We spend about the same amount of time earning the money to pay for our car as we do driving it. Whodathunkit?

Now we can calculate our average driving speed. The total time we devote to driving (and we are certainly devoted to our cars) is 3000 hours driving plus 3000 hours earning, which equals 6000 hours. Thus, our average automotive speed is

$$s = \frac{120,000\,mil}{6000\,hr} = 20\,mph$$

And this does not even count the time spent looking for a parking space or waiting for the mechanic.

---

* Not counting taxes.

Energy and Work

Chapter 6

* * * * * * * * * * * * * * *

6.1 Energy of Height

Gravity sucks! If you drop something, it will fall. The gravitational acceleration at the surface of the Earth is $g = 10 \, \text{m/s}^2$.* This means that a falling object will increase its speed by 10 m/s (about 20 mph or 36 kph) every second. If you fell for 5 s, you would hit the ground at 50 m/s or 110 mph or 180 kph. Ouch!

g is also the gravitational force (measured in newtons (N)) exerted on an object of mass 1 kg (1 kg is about the mass of 2 lb) at the surface of the Earth. Thus, a 1-kg block experiences a gravitational force (i.e., has a weight) of 10 N on Earth (less on the moon and more on the "surface" of Jupiter). In general, the gravitational force on an object can be expressed as $F = mg$, where m is the mass in kilograms.†

It takes work to lift an object against the force of Earth's gravity. The energy needed to do this is called the *potential energy*

$$\text{PE} = mgh$$

where h is the height in meters. This makes sense. If you increase the mass of the lifted object, or the height you lift it to, or the gravitational pull, then you will need more energy to lift the object. The metric unit of energy is the joule, abbreviated J.

---

* To be precise, which this book is not, g varies from 9.78 to 9.83 m/s² depending on latitude.

† This difference between mass and weight can be confusing. The object has the same mass everywhere, but the force needed to lift it depends on the planetary gravity. To increase the potential confusion, in US customary units, pounds can refer to either weight or mass.

Mountain climbing

6.1.1

How much do you change your potential energy climbing a medium-sized mountain? How does this compare to the 6×10^5 J in a can of soda?

* * * * * * * * * * * * *

HINT: Divide your weight in pounds by two to get your approximate mass in kilograms.

HINT: Mt. Everest, the tallest mountain on Earth, is 10^4 m (30,000 ft).

ANSWER: Since potential energy is PE $= mgh$, we need to estimate our mass, the gravitational acceleration, and the height of the mountain. Assuming that we do this on Earth, $g = 10 \, \text{m/s}^2$.

Humans (at least the ones likely to climb mountains) weigh between 100 and 200 lb. This means that we have a mass between 50 and 100 kg. As usual, we will choose 100 kg because it is a round number* and makes the arithmetic easier.

A medium-sized mountain is taller than a building and shorter than Mt. Everest. The tallest building has about 100 floors or stories with about 10 ft per floor and thus is about 1000 ft tall. Since $1 \, \text{m} \approx 3 \, \text{ft}$, this is about 300 m. Thus, a medium-sized mountain is somewhere between 3×10^2 and 1×10^4 m. We'll take the average of the coefficients (the average of 3 and 1 is 2) and the average of the exponents (the average of 2 and 4 is 3) to get a height of $h = 2 \times 10^3$ m (or 6000 ft). This is the height of the tallest mountain on the US east coast. If you chose a shorter or taller mountain, that's fine too.

Now we can calculate the change in potential energy when you climb that mountain:

$$\text{PE} = mgh$$
$$= 100 \, \text{kg} \times 10 \, \text{m/s}^2 \times 2 \times 10^3 \, \text{m}$$
$$= 2 \times 10^6 \, \text{J}$$

Gee, two million joules! Is a joule big or small? Is that a lot?

To figure that out, we need to compare to other measures of energy. A 12-oz (330-mL) can of soda contains 6×10^5 J of food energy.† Thus, by climbing

---

* And if I massed a 100 kg, I would be round too.

† If you live in the US, the soda's energy content is given in Calories. We'll discuss those in the next chapter. If you live elsewhere, it's given directly in joules.

that mountain, we would gain potential energy equal to the food energy contained in

$$N_{\text{sodas}} = \frac{2 \times 10^6 \, \text{J}}{6 \times 10^5 \, \text{J/soda}} = 3 \text{ sodas}$$

That's not much! Climbing a 2-km (6000-ft) mountain takes a lot more work than that!

Flattening the Alps

6.1.2

How much energy could we get from flattening the Rocky Mountains or the Alps (i.e., how much potential energy is stored in a mountain range)?

* * * * * * * * * * * * * *

HINT: What is the average height of the Alps or the Rockies?

HINT: How long is the mountain range and how wide is it?

HINT: What is the average density of rock?

HINT: The average density of rock is more than water (1 ton/m³) and less than iron (10 tons/m³).

ANSWER: We need to estimate the average height of the Alps or the Rockies and their total mass. To get the mass we will need to estimate the volume and the density. To get the volume we will estimate the average length, width, and height. The highest mountain in either mountain range is about 20,000 ft (6000 m) so we will estimate that the average height including all the peaks and valleys is less than half of that or 2×10^3 m.*

The Alps stretch about 10^3 (1000) km from east to west (it is certainly more than 10^2 km and less than 10^4 km) and about 200 km from north to south. (If you did this in miles, that's fine. Since miles and kilometers are so close, 1 mi = 1.6 km, they are almost interchangeable in this book.) Now we need to convert these distances to meters. Fortunately, it is easier than converting miles to feet. Since 1 km = 10^3 m, 10^3 km = 10^6 m and 2×10^2 km = 2×10^5 m. Thus, the volume of the Alps will be

$$V_{\text{Alps}} = l \times w \times h = 10^6 \, \text{m} \times 2 \times 10^5 \, \text{m} \times 2 \times 10^3 \, \text{m}$$
$$= 4 \times 10^{14} \, \text{m}^3$$

The Rockies are a lot longer since they run from southern US to northern Canada. This is about 3000 miles or 5×10^3 km. That would increase the volume by a factor of five from 4×10^{14} m^3 to 2×10^{15} m^3.

Now we estimate the density in order to convert volume to mass. Mountains are made of rock. Rock sinks and thus is denser than water (1 ton/m^3). Rock is less dense than iron (10 ton/m^3). Thus, we will use a

---

* If you look at a relief map of the Alps, you will see that most of the Alps are actually only at about 1500 m. Thus, our crude estimate is not too bad.

density of $d = 3\,\text{tons/m}^3$ or $3 \times 10^3\,\text{kg/m}^3$. This gives a total mass of

$$M_{\text{Alps}} = V \times d = 4 \times 10^{14}\,\text{m}^3 \times 3 \times 10^3\,\text{kg/m}^3$$
$$= 1 \times 10^{18}\,\text{kg}$$

and the Rockies will have five times as much. That's a lot of mass.

Now we can calculate the potential energy. Note that the average height of the mountain range is 2×10^3 m. However, the mass extends from sea level up to the average height. Thus, for calculating the potential energy, we need to use half of that average height.*

$$\text{PE}_{\text{Alps}} = mgh = 1 \times 10^{18}\,\text{kg} \times 10\,\text{m/s}^2 \times 1 \times 10^3\,\text{m}$$
$$= 1 \times 10^{22}\,\text{J}$$

and the Rockies will have five times as much or $\text{PE}_{\text{Rockies}} = 5 \times 10^{22}$ J (about 10^{17} cans of soda). Now THAT is a LOT of energy! The energy to raise the mountains came from the motion of the continents (and originally from the heat generated in the core of the Earth).

Note that the height of the mountain range enters our equations twice, once explicitly as the height and once as part of the volume. This means that if we overestimate the height by a factor of two, we will overestimate the potential energy by a factor of four. Fortunately, we are only trying to get within a factor of ten.

---

* It's OK if you left out this factor of two. We are only trying to get within a factor of ten of the correct answer.

Raising a building

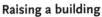

6.1.3

How much potential energy
does a 100-story building
have?

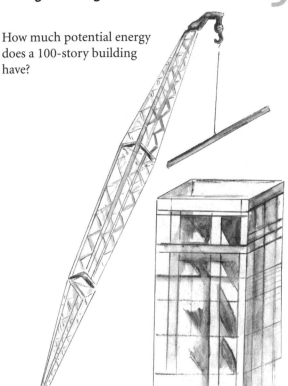

✻ ✻ ✻ ✻ ✻ ✻ ✻ ✻ ✻ ✻ ✻ ✻ ✻ ✻ ✻

HINT: What fraction of the building's volume is steel structure?

HINT: The building is made mostly of steel
(density ≈ 10 tons/m³).

HINT: What is the length and width of the building?

HINT: How tall is each story?

HINT: How tall is the building?

ANSWER: As in the previous problem, we need to estimate the height of the object and its mass.

Each story of a building is about 10 to 12 feet. Ceilings are about 8 feet (i.e., there is about 8 feet of air between the floor and the ceiling) and there is space used for ductwork and cabling and for structure (holding up the building). This 10 to 12 feet is equal to 3 to 4 meters. Thus, the 100 story building is 100 stories × 3 m/story = 300 m tall. The average height of the material in the building will be half of that (since not all the material is located at the top of the building) or $h = 150$ m.

Now we need to estimate the mass of the building. The building's mass is in its vertical structure and its horizontal floors. The structural supports are all steel (since it is much stronger than the same weight of concrete). The structural supports will occupy much more than 1% and much less than 100% of the volume of the building. We will take the geometric mean and estimate that they occupy about 10% of the volume of the building (if it was much less then we could build much taller buildings and if it was much more then there would be little usable space in the building).

Therefore, we need to estimate the volume of the building. We already have the height. The area of a typical 100-story building will be somewhere between a football field (50 yards by 100 yards = 5×10^3 yd^2 = 5×10^3 m^2) and a private house (10^3 ft^2 or 100 m^2). We'll use 10^3 m^2 for our building (this is 30 m by 30 m or 100 ft by 100 ft). Thus, the volume is

$$V_{\text{bldg}} = h \times A = 3 \times 10^2 \, \text{m} \times 10^3 \, \text{m}^2$$
$$= 3 \times 10^5 \, \text{m}^3$$

We'll assume that 10% of this is steel with a density $d = 10 \, \text{tons/m}^3 = 10^4 \, \text{kg/m}^3$. Thus, the mass of the building is

$$m = V_{steel} \times d = V_{bldg} \times 10\% \times d$$
$$= 3 \times 10^5 \, m^3 \times 0.1 \times 10^4 \, kg/m^3$$
$$= 3 \times 10^8 \, kg$$

Note that the 110-story Sears Tower has a mass of $M = 2 \times 10^8 \, kg$ (so we are gol-durn close!).

Now the potential energy is easy:

$$PE = mgh = 3 \times 10^8 \, kg \times 10 \, m/s^2 \times 1.5 \times 10^2 \, m$$
$$= 5 \times 10^{12} \, J$$

That's equivalent to the energy contained in 10 million cans of soda. Now that is a lot of energy. It helps explain why it is much easier to collapse a building than it is to build it.

6.2 Energy of Motion

It takes work to make an object change its speed (e.g., from 0 to 60 mph) even ignoring pesky details such as friction and air resistance. The kinetic energy (energy of motion) of an object (in joules) is

$$KE = \frac{1}{2}mv^2$$

where m is the mass in kilograms and v is the velocity (or speed) in meters per second.* (It is easy to convert from meters per second to miles per hour since 1 m/s $\approx$ 2 mph.)

If you drop a water balloon from a building, it will start with a large amount of potential energy (PE), convert PE to KE as it falls, and then convert KE to a large splash as it hits.

---

* Technically, velocity includes both magnitude and direction, while speed includes just the magnitude. Since we will only be using them to calculate kinetic energy where the direction does not matter, we will use both terms interchangeably.

At your service

What is the kinetic energy in joules of a served tennis ball?

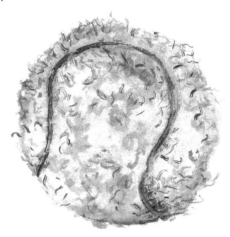

HINT: What is the weight of a tennis ball?

HINT: How many tennis balls are needed to weigh a pound (or a kilogram)? One, ten, one hundred?

HINT: What speed does the tennis ball have? 1 mph, 10 mph, 100 mph, . . .

HINT: Remember that 1 kg ≈ 2 lb and 1 m/s ≈ 2 mph.

ANSWER: Since $KE = \frac{1}{2}mv^2$, we need to estimate the mass and the velocity. There are definitely more than one and less than 100 tennis balls in a pound so we will estimate that there are 10 tennis balls per pound. This means that there are 20 tennis balls in a kilogram (since $1\,kg \approx 2\,lb$). Thus, the mass is

$$m = \frac{1\,kg}{20\,balls} = 5 \times 10^{-2}\,kg/ball$$

Now we need to estimate the velocity. We can place certain limits on the speed. Since the tennis ball does not make a sonic boom, we know it is slower than the speed of sound (300 m/s). The tennis ball travels faster than a car (30 to 60 mph or 15 to 30 m/s). Let's see if we can get more information. We can compare it to baseball. A pitched fastball (in baseball) travels at 100 mph (50 m/s). A served tennis ball should go faster than that because the racket gives some mechanical advantage. Alternatively, we can use the size of the tennis court. The ball travels about 60 ft (20 m) from the service line to where it hits the court. It must take more than 0.2 s to do this (or we could never return a serve) and it certainly takes less than a second. This gives a range of speeds from 20 to 100 m/s. We'll use 60 m/s (or 120 mph).

Now we can calculate the kinetic energy:

$$KE = \frac{1}{2}mv^2 = 0.5 \times 5 \times 10^{-2}\,kg \times (60\,m/s)^2$$
$$= 100\,J$$

100 J is the energy output of a 100-W light bulb for 1 s (since 1 watt = 1 joule/second).

Not much!

Now let's compare to reality. The fastest tennis serve was 164 mph (73 m/s) so 60 m/s is quite reasonable for an excellent player. The mass of a tennis ball is 57 g, so our estimate of 50 g is also very close.

Kinetic trucking

6.2.2

What is the kinetic energy (in joules) of a large truck at highway speed?

* * * * * * * * * * * * * * * *

HINT: A small car has a mass of about 1 ton.

HINT: What is its mass?

HINT: Remember that 2 mph ≈ 1 m/s.

HINT: What is its speed?

ANSWER: We need to estimate the speed and mass of the truck. US highway speed limits are typically between 55 and 65 mph. We'll use a speed of 60 mph ≈ 30 m/s. (Remember that we need to use metric units in our equations.) A small car weighs about 1 ton. A large truck will weigh a lot more than one car but less than 100 cars. This would give us a weight of 10 tons (the geometric mean of 1 and 100). Alternatively, we can estimate the mass from bridge weight-limit signs. Bridge weight limits, if posted, tend to be about 10 or 20 tons. This implies that large trucks weigh more than that (since otherwise the warning signs would not be needed). This gives a weight of about 40 tons. We'll use the average of 10 tons and 40 tons, or 20 tons. Since 1 ton = 10^3 kg, 20 tons = 20×10^3 kg = 2×10^4 kg. Now we calculate that

$$\mathrm{PE} = \frac{1}{2}mv^2 = 0.5 \times 2 \times 10^4 \,\mathrm{kg} \times (30 \,\mathrm{m/s})^2$$
$$= 1 \times 10^7 \,\mathrm{J}$$

This is 10^6 (one million) times less than the potential energy of a large building. Since a typical can of soda contains 10^5 J, this is the energy contained in 100 12-ounce (330-mL) cans of soda. We'll investigate automotive energy sources (batteries and gasoline) more in the next chapter.

Note that a small car with a mass of about 1 ton will have a lot less kinetic energy. We can calculate it directly, or we can use the fact that the car has a mass that is twenty times less than the truck and therefore will have a kinetic energy that is twenty times less. Thus, the car's kinetic energy will be $\mathrm{KE}_{car} = 1 \times 10^7 \,\mathrm{J}/20 = 5 \times 10^5$ J. This is the energy contained in five cans of soda.

What is the kinetic energy of a drifting continent?
(Ignore the rotation and other motion of the Earth.
We are only interested in the motion of the continent
with respect to the rest of the Earth.)

✳ ✳ ✳ ✳ ✳ ✳ ✳ ✳ ✳ ✳ ✳ ✳ ✳ ✳ ✳ ✳

HINT: What is the volume of the continent?

HINT: The circumference of the Earth is 4×10^4 km.

HINT: The depth of the continent is only about 50 km.

HINT: What is the speed of the continent?

HINT: It took 10^8 years for the Atlantic Ocean to be formed by
North America moving away from Europe.

HINT: The Atlantic Ocean is 5000 km wide at its widest point.

HINT: One year $= \pi \times 10^7$ s.

HINT: The density of rock is somewhere between water
(10^3 kg/m³) and iron (10^4 kg/m³).

ANSWER: As usual, we need to estimate the mass and the speed. Let's start with the mass. We will estimate the volume (i.e., length, width, and height [or depth]) and the density using North America. We estimated the width of the contiguous US back in question 3.9 as 1/8 of the Earth's circumference or 5×10^6 m. Assuming that North America is square (insert your bad joke here; we're temporarily out of stock), the length is also 5×10^6 m.

The depth of the crust is a little more difficult. It is certainly more than 1 km (many mines go deeper than that) and less than 10^3 km (since we know that the tectonic plates are a small fraction of the radius of the Earth). We'll take the geometric mean of 1 and 10^3 km which is 30 km or 3×10^4 m.* Thus, the volume of the North American plate is

$$V_{NA} = lwh = 5 \times 10^6 \text{ m} \times 5 \times 10^6 \text{ m} \times 3 \times 10^4 \text{ m}$$
$$= 8 \times 10^{17} \text{ m}^3$$

Now we can multiply that by the density of rock to get the mass. The density must be more than that of water ($d = 1 \times 10^3$ kg/m^3) and less than that of iron ($d = 10 \times 10^3$ kg/m^3) so we will use $d = 3 \times 10^3$ kg/m^3:

$$m_{NA} = d \times V = 3 \times 10^3 \text{ kg/m}^3 \times 8 \times 10^{17} \text{ m}^3$$
$$= 2 \times 10^{21} \text{ kg}$$

Now we need the velocity. If you read about California earthquakes, you might know that the North American plate is moving at about 1 to 2 cm per year. Alternatively, we can use a longer time scale. The Atlantic Ocean has been widening for about 10^8 years as North America has been moving away from Europe. At present, the maximum width of the Atlantic Ocean

---

* The actual thickness of the continental crust is between 20 and 80 km.

is about $5000 \, \text{km} = 5 \times 10^6 \, \text{m}$. Thus, the velocity* of North America is

$$v_{NA} = \frac{d}{t} = \frac{5 \times 10^6 \, \text{m}}{10^8 \, \text{yr}}$$

$$= 5 \times 10^{-2} \, \text{m/yr} \times \frac{1 \, \text{yr}}{\pi \times 10^7 \, \text{s}}$$

$$= 2 \times 10^{-9} \, \text{m/s}$$

This estimate of 5 cm per year is a few times larger than the actual speed. It is still not very fast.

The kinetic energy is now

$$KE = \frac{1}{2} m v^2 = 0.5 \times 2 \times 10^{21} \, \text{kg} \times (2 \times 10^{-9} \, \text{m/s})^2$$

$$= 4 \times 10^3 \, \text{J}$$

That is a lot more than the tennis ball but a lot less than the truck! It is much less than one can of soda.

To explain why continents are so hard to stop, we would need to estimate their momentum. Perhaps in the next book . . .

---

* We use V for volume and v for velocity. Scientists rapidly run out of variables and have to reuse letters.

"To boldly go ..."

6.2.4

How much energy is needed to get a spaceship
from Earth to Alpha Centauri (the nearest star,
about 4 light-years away) before
the passengers die of old age?
How many tons of fuel will this
take (assuming 4×10^9 J/ton
of TNT or rocket fuel)?

※ ※ ※ ※ ※ ※ ※ ※ ※ ※ ※ ※ ※ ※ ※ ※

HINT: How large a ship do you want?

HINT: An aircraft carrier is 10^5 tons.

HINT: The speed of light is $c = 3 \times 10^8$ m/s. A light-year is the
distance light travels in one year (or $\pi \times 10^7$ s).

HINT: What velocity would the spaceship need to have?

HINT: If light makes the journey to Alpha Centauri in 4 years,
what fraction of light speed would the ship need to get there with
living passengers?

HINT: How much kinetic energy would the spaceship have at
that velocity?

135

ANSWER: The rocket fuel will provide the kinetic energy for the rocket ship. To figure out the kinetic energy, we need to estimate the mass and the velocity. Thus, we first need to figure out how fast the spaceship will need to travel. The spaceship needs to travel the 4 light-years from Earth to Alpha Centauri in 40 years or less.[*] Since it takes us 40 years to travel the distance that light travels in four, our velocity is one-tenth of the speed of light. Thus,

$$v = \frac{1}{10}c = 0.1 \times 3 \times 10^8 \text{ m/s}$$
$$= 3 \times 10^7 \text{ m/s}$$

Now we need to estimate the mass of the rocketship. A modern aircraft carrier displaces 10^5 tons. Columbus sailed to the New World in the *Santa Maria*, which weighed only about 100 tons.[†] It's hard to imagine cramming the life-support systems needed for a 40-year voyage in a tiny ship. Let's use a mass of 10^4 tons or 10^7 kg. Since the kinetic energy is directly proportional to the mass, it's easy to try different masses. Anyway, let's calculate the fuel we need.

The spaceship will have a kinetic energy of

$$\text{KE} = \frac{1}{2}mv^2 = 0.5 \times 10^7 \text{ kg} \times (3 \times 10^7 \text{ m/s})^2$$
$$= 5 \times 10^{21} \text{ J}$$

That's a lot of energy. We could get that much from flattening the Alps, but that would supply the energy for only one small vessel.

---

[*] Why 40 years? We could have chosen any number between 20 and 60 years. There are two reasons: (1) it worked for Moses and (2) it simplifies the arithmetic because it is exactly ten times four.
[†] But his voyage lasted only a few months and he did not have to bring all his oxygen.

Now let's look at the fuel needed. At 4×10^9 J/ton of fuel, our 10^4-ton spaceship will need fuel with a total mass

$$m_{\text{fuel}} = \frac{5 \times 10^{21} \text{ J}}{4 \times 10^9 \text{ J/ton}} = 1 \times 10^{12} \text{ ton}$$

Yikes! It does not matter what size spaceship you chose because the fuel will weigh 100 million (10^8) times more than the spaceship. Now we need to include the extra fuel needed to accelerate the fuel. We did not even take into account the energy needed to decelerate the spaceship at its destination or the inefficiency of rocket engines.

There is no way we can do this. We cannot fly a spaceship to the stars in a human lifetime using conventional chemical fuels.

Someone needs to start working on dilithium crystals or antimatter drives.

6.3 Work

"Work" is not just a generic term. In physics it means transferring energy by pushing on something (applying a force) and making it move. When you kick a football (either the spherical or nonspherical kind), you are applying a force over a distance. When you shoot a bullet from a gun, the expanding gases in the gun barrel exert a force over a distance. When you apply the brakes in your car, the ground applies a force on your car over a distance. The energy transferred by this force is

$$W = Fd$$

where F is the force in newtons ($10\,\mathrm{N}$ is the weight of $1\,\mathrm{kg}$ and $10^4\,\mathrm{N}$ is the weight of 1 ton [on Earth]) and d is the distance over which the force is applied (in meters). Note that the force must be in the same (or opposite) direction as the distance traveled. If you push in the same direction that the object travels, you increase its energy and speed (e.g., kicking a football). If you push in the opposite direction, then you decrease the object's energy and speed (e.g., apply the brakes).* Thus,

work = change in kinetic energy

---

* If you push sideways (perpendicular to the object's motion), then you do no work. You change the object's direction but not its energy or speed (e.g., the Earth pulling on the Moon).

Crash!

6.3.1

In the interests of researching this book, John offers to crash his car into a hard barrier at highway speed. What force would be exerted on his body as he stopped? (Assume that he is wearing his seat belt and his air bag deploys properly.)

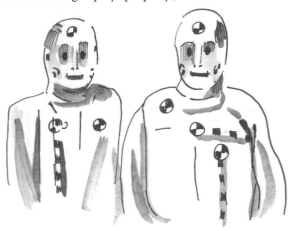

HINT: The car crumples about 0.5 m (between 1 and 2 ft).

HINT: The distance involved is the amount that the front end of his car crumples during the crash.

HINT: The seat belt and air bag hold him in his seat so that he decelerates at the same rate as his car.

HINT: Estimate his mass and speed.

HINT: What is John's kinetic energy before the crash?

HINT: You will need to estimate his change in kinetic energy and the distance he travels during the crash.

ANSWER: For John to change his velocity from high-way speed to a complete stop, a force needs to be exerted on him. This is Newton's first law: Your speed and direction of motion will not change unless an outside force acts on you. In this case, the force acting on John is exerted by the seat belt and air bag that compel him to slow down as the car slows down. We will estimate the force exerted on him by the seat belt and air bag. To do this, we will estimate his initial kinetic energy (since the work done by the seat belt must change that value to zero) and the distance over which the force acts.

We'll assume that he is driving at 60 mph or 90 kph (30 m/s) and has a mass of 100 kg (220 lb).* In that case, his kinetic energy is

$$\text{KE} = \frac{1}{2}mv^2 = 0.5 \times 100 \,\text{kg} \times (30 \,\text{m/s})^2$$
$$= 5 \times 10^4 \,\text{J}$$

Thus, the work done in stopping him must be 5×10^4 J.

Fortunately, the front end of his car is well designed and crumples. While the front bumper stops imme-diately, the passenger compartment slows and stops more gradually as the front end crumples. A typical front end will crumple less than 1 m (3 ft) and more than 0.1 m (10 cm or 4 in.). We'll take the average and use 0.5 m.† This means that the force exerted is

$$F_{\text{crash}} = \frac{W}{d} = \frac{5 \times 10^4 \,\text{J}}{0.5 \,\text{m}}$$
$$= 10^5 \,\text{N}$$

---

* This mass is in the cause of science. He will lose the extra weight when the problem is finished.

† In this case, there is little difference between the geometric mean of 0.3 m and the average of 0.5 m.

Let's see how much that is. Since 10 N is the gravitational force exerted on a 1-kg object, 10^5 N is the gravitational force exerted on a 10^4-kg object. Thus, the force exerted on John during the collision is equivalent to one hundred 100-kg (200-lb) people standing on your chest. That is 10 tons! Yikes! Driving into a wall at 60 mph (even if you are wearing your seat belt) is *not* a good idea!

Car dashboards are padded and cars have crumple zones to increase the stopping distance in a crash in order to decrease the stopping forces involved. The reason that you should always wear your seat belt is that the seat belt holds you in your seat, so that you stop gradually as the car stops. If you are not wearing your seatbelt, your head will stop abruptly as it hits the windshield. Since the stopping distance will be much less, the forces involved will be much greater. This is generally not a good thing.

Spider-Man and the subway car 6.3.2

In the movie *Spider-Man 2*, Spider-Man stops a runaway New York City six-car subway train by attaching his webs to nearby buildings and pulling really hard for 10 or 20 city blocks. How much force does he have to exert to stop the subway train? Give your answer in newtons and in tons (1 ton = 10^4 N). How does this compare to the force that you can exert?

✳ ✳ ✳ ✳ ✳ ✳ ✳ ✳ ✳ ✳ ✳ ✳ ✳ ✳ ✳

HINT: The stopping distance is about 1 km.

HINT: How much work is needed to stop the train? (Remember that work = change in kinetic energy or $W = \Delta KE$.)

HINT: What are its mass and velocity?

HINT: What is the kinetic energy of a subway train at normal speed?

ANSWER: This is very similar to the previous question. Since the work done by Spider-Man to stop the train is equal to the train's initial kinetic energy, we need to estimate the mass and velocity of the train. We will then need to estimate the stopping distance in order to calculate the force exerted.

A subway car is about the same size and weight as a semi-trailer (18-wheeler) truck. This is between 10 and 40 tons. We'll use 20 tons (or 2×10^4 kg). There are six cars on a train so that the mass of the train is $6 \times 2 \times 10^4$ kg $= 10^5$ kg. They certainly go faster than 20 mph and slower than 100 mph. Since it is not that far between subway stops, subways travel at only about 40 mph (20 m/s). Thus, the kinetic energy of a subway train is

$$\text{KE} = \frac{1}{2}mv^2 = 0.5 \times 10^5 \text{ kg} \times (20 \text{ m/s})^2$$
$$= 2 \times 10^7 \text{ J}$$

Now we need to figure out the stopping distance. There are 20 blocks per mile in Manhattan. Thus, 10 or 20 blocks is about 1 km or 10^3 m. (It's certainly more than 100 m and less than 10 km.) Thus, Spider-Man needs to exert a force

$$F = \frac{\text{KE}}{d} = \frac{2 \times 10^7 \text{ J}}{10^3 \text{ m}}$$
$$= 2 \times 10^4 \text{ N}$$

A force of 2×10^4 N is the weight of 2000 kg or 2 tons. For a superhero who can lift cars, this is quite possible (although definitely not easy). A human could definitely not do it.

Wow! Hollywood got the physics correct, in a superhero movie no less! Hurray!

Hydrocarbons and Carbohydrates

Chapter 7

✳ ✳ ✳ ✳ ✳ ✳ ✳ ✳ ✳ ✳ ✳ ✳ ✳ ✳ ✳

7.1 Chemical Energy

Unless we are capable of photosynthesis, we get most of our energy from chemical reactions: from eating food and from burning hydrocarbon fuels. In a typical chemical reaction, one electron is exchanged between two atoms. The energy of this exchange is about

1.5 electron volts or 1.5 eV.* If you want more precision than that, ask a chemist or look it up. To convert this to a useful number, we need to know two things:

1. The conversion from electron volts to joules:
   $1 \text{ eV} \approx 2 \times 10^{-19} \text{ J}$

2. The number of molecules involved in the reaction

To determine the second, we need to introduce a little chemistry. We will be concerned primarily with hydrocarbons and therefore will limit ourselves to the reactions $C + O \rightarrow CO_2$ (carbon plus oxygen reacts to form carbon dioxide) and $H + O \rightarrow H_2O$ (hydrogen plus oxygen reacts to form water). All the oxygen for these reactions comes from the atmosphere.

In this book, there are only three hydrocarbons: coal (pure carbon), natural gas (methane or CH_4), and everything in between (including gasoline), which we will call CH_2.† One mole of carbon has a mass of 12 g (the atomic weight of carbon) and contains $N_A = 6 \times 10^{23}$ atoms. Burning that will result in N_A chemical reactions. One mole of methane has a mass of 16 g (the atomic weight of the carbon plus four hydrogens) and contains $N_A = 6 \times 10^{23}$ molecules. Burning that will result in $3 \times N_A$ reactions (one CO_2 and two H_2O).

---

* We know this because batteries convert chemical energy to electrical energy. Common batteries provide an electrical potential of 1.5 V. Therefore, each single electron flowing through the battery gains an energy of 1.5 electron volts and each coulomb of electricity (a coulomb is a LOT of electrons, $1 \text{ C} = 6 \times 10^{18} e$) flowing through the battery gains an energy of 1.5 coulomb volts (or 1.5 joules).

† This is because many hydrocarbons consist of long carbon chains with two hydrogen atoms per carbon atom. We can think of these as being made up of repeated CH_2 units. In practice, the ratio of hydrogen to carbon varies depending on the exact chemical involved. We are not chemists; this is not our problem.

Energy in gasoline

7.1.1

How much chemical energy (in joules) can be released by burning 1 kg (about 1 L or 1/4 gal) of gasoline? What is its energy density (in J/kg)?

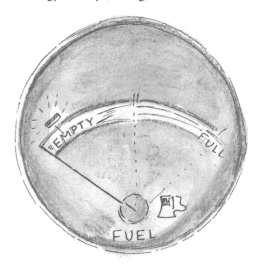

HINT: Assume that gasoline is approximately CH_2.

HINT: How many moles of CH_2 are in 1 kg of gasoline?

HINT: One mole of CH_2 has a mass of 14 g.

HINT: We can get 3 eV from burning one molecule of CH_2 (since we make one CO_2 and one H_2O molecule).

ANSWER: Since we know that the energy we get for each chemical reaction is 1.5 eV, we need to estimate the number of chemical reactions that occur when we burn 1 kg of gasoline. To do this we need to estimate the chemical composition.

We will assume that the hydrogen to carbon ratio in gasoline is two (since it is more than zero [pure carbon] and less than four [pure methane]). Thus, we will assume that gasoline is made of CH_2 molecules.* The atomic masses of carbon and hydrogen are 12 and 1, respectively, so that the molecular mass of CH_2 is 14. This means that one mole of CH_2 has a mass of 14 g or 1.4×10^{-2} kg. Thus, 1 kg of gasoline contains

$$N = \frac{1 \text{ kg}}{1.4 \times 10^{-2} \text{ kg/mole}} = 70 \text{ moles}$$

Each of these CH_2 "molecules" will give us two reactions: the carbon atom will oxidize and form CO_2 and the two hydrogen atoms will oxidize and form H_2O. Thus, each CH_2 molecule will provide 3 eV. The total energy released by burning (oxidizing) 1 kg of gasoline will be

$$E = \frac{70 \text{ moles}}{\text{kg}} \times \frac{6 \times 10^{23} \text{ reactions}}{\text{mole}} \times \frac{3 \text{ eV}}{\text{reaction}}$$

$$\times \frac{1 \text{ J}}{6 \times 10^{18} \text{ eV}} = 2 \times 10^7 \text{ J/kg}$$

Thus, we estimate that 1 kg of gasoline will release 2×10^7 J when burned.

Looking this up on the web [15], we find that gasoline has an energy density of about 4.5×10^7 J/kg so we are only off by a factor of two. Not bad, considering the approximations we made.

---

* This ignores the fact that the carbon atoms are in long-chain molecules (octane has 8 carbon atoms in a chain) and thus ignores the energy needed to break the carbon–carbon bonds. We'll let the chemists worry about that.

Note also that gasoline has a density of about 3/4 that of water. This is definitely close enough to one for this book. However, if you need to be precise,* you should use a volume energy density of 3×10^7 J/L.

Note that 1 kg of TNT contains only 4×10^6 J, which is only 10% of gasoline. However, the TNT can release that energy MUCH more rapidly.

* Which this book is not.

Battery energy

How much chemical energy is stored
in a common D-battery?

HINT: The energy used (in J) equals the power output (in watts)
times the lifetime in seconds.

HINT: Compare the light output of a D-battery-powered
(non-LED) flashlight with a 100-W light bulb or with a 4-W
night-light.

HINT: How long will a battery last in that flashlight?

ANSWER: To estimate the chemical energy contained in a battery, we need to estimate its power output and its lifetime. This is because power (measured in watts) is the energy used each second. Thus, a 100-W light-bulb consumes 100 J of electrical energy every second.

The easiest way to estimate the power output of a battery is to compare the light output of a flashlight powered by the battery with other lights of known power consumption. Since different types of light bulb (e.g.; incandescent, fluorescent, LED) have very different efficiencies (i.e., very different amounts of light produced for the same amount of energy consumed), we need to make sure we compare light from the same type of bulb. We'll use the common incandescent bulbs for our comparison. (Those are the ones that emit light by heating the filament to thousands of degrees and hence get very hot.)

A flashlight definitely gives off much less light than standard lamp bulbs (100-W, 60-W, or 25-W). It gives off about the same amount of light (within a factor of 10) as a 4-W night-light, although it is difficult to compare precisely because a flashlight is usually directional (i.e., it uses a mirror to focus the beam in one direction) and a night-light is usually omni-directional and partly shielded.* Thus, we will estimate that the flashlight consumes 4 W (i.e., 4 J/s).

Now we need to estimate the lifetime of the battery. The flashlight can stay lit for more than an hour but less than a day (24 hr). We will take the geometric mean and estimate that the lifetime is 5 hours. Now

---

* You also need to compare the two light sources under similar ambient light conditions (since the light sensitivity of the human eye changes by orders of magnitude from bright sunlight to darkness).

we can calculate the chemical energy stored in the battery:

$$E_{D-battery} = \text{power} \times \text{time}$$
$$= 4\,W \times 5\,hr \times 3.6 \times 10^3\,s/hr$$
$$= 7 \times 10^4\,J$$

Thus, one D-battery contains $7 \times 10^4\,J$ of chemical energy.

Let's compare our estimates to reality. A Duracell alkaline D-battery has a specified capacity of 15,000 mA-hr (milliamp-hours) or 15 A-hr. The power output equals the current (in amperes) times the voltage. Thus, we have

$$E_{spec} = 15\,A\text{-hr} \times 1.5\,V \times 3.6 \times 10^3\,s/hr$$
$$= 8.1 \times 10^4\,J$$

Now we need to ask, is this a lot of energy? It is about the same amount as the chemical energy contained in a can of soda. It is also about 1/1000 of the energy in a kilogram (or liter) of gasoline. Since $10^{-3}\,L = 1\,mL = 1\,cm^3$, it is the energy contained in a cubic centimeter (about the volume of the last joint your smallest finger) of gasoline. Not much.

We'll explore this further in the next question.

What is the energy density of a D-battery (in joules per kilogram)? How does this compare to the energy density of gasoline?

HINT: How many D-batteries are there in a kilogram or a pound? 1? 10? 100? 1000?

ANSWER: To estimate the energy density, we need to estimate the energy and mass of one battery. We estimated the energy of one D-battery in the previous problem. Now we just need to estimate the mass. It is frequently easier to estimate how many objects it takes to have the weight of 1 kg rather than to estimate the weight of a single object. One D-battery clearly weighs less than a kilogram and 100 D-batteries clearly weigh more. We estimate that there are about 10 D-batteries per kilogram. (In reality, a four-pack of D-batteries has a shipping weight of one pound. This would imply exactly nine batteries in a kilogram.)

Now we can calculate the energy density of alkaline D-batteries:

$$\text{energy density}_{\text{D-cell}} = \frac{\text{energy}}{\text{battery}} \times \frac{\text{batteries}}{\text{kg}}$$

$$= \frac{8 \times 10^4 \text{ J}}{\text{D} - \text{cell}} \times \frac{10 \text{ D} - \text{cell}}{\text{kg}}$$

$$= 8 \times 10^5 \text{ J/kg}$$

Remember that gasoline has an energy density of 4×10^7 J/kg. This is a factor of fifty (50) worse. Wow!

In fact, reusable batteries are much worse than that. Reusable batteries are designed for, among other things, energy density and cycle life (how many times you can recharge it) [16]. The energy densities of rechargeable batteries are several times worse than the nonrechargeable ones. They range from 1×10^5 J/kg for lead-acid batteries to 6×10^5 J/kg for the highest energy-density lithium-ion batteries. The batteries can be recharged only 200 to 500 times (depending on type of battery).

Note that the cycle life can also be a serious limitation. If you refuel your car once per week, that is

50 times per year. This means that a 200-cycle battery would last only four years as the energy storage for a car. If you are buying an electric car, you do not want to have to replace the large, heavy, expensive battery every four years.

Batteries vs. gas tanks

7.1.4

How many tons of batteries will you
need to contain the same amount of
energy as the gasoline in your gas
tank? How much does a full tank
of gasoline weigh?

* * * * * * * * * * * * * * *

HINT: How many gallons of gas does it take to refuel your car
when the tank is almost empty? (Remember that 1 gal = 4 L.)

HINT: At 3×10^7 J/L, how much energy does your gas tank
contain?

ANSWER: Automobile (passenger car and SUV) gas tanks range from 10 to 30 gallons depending on the size and mileage of the car. A compact car takes about 10 gallons, a minivan takes about 20, and a Hummer H2 takes 32. Let's use 20 gallons as a reasonable average. Twenty gallons of gas contains energy

$$E = 20\,\text{gal} \times 4\,\text{L/gal} \times 3 \times 10^7\,\text{J/L}$$

$$= 2 \times 10^9\,\text{J}$$

or 2 GJ. Twenty gallons is about 80 L and therefore has a mass of about 80 kg or 160 lb.

Now let's see how large a rechargeable battery we need to store that much energy. Let's use the best lithium-batteries available in 2006, with an energy density of 6×10^5 J/kg. We will need a battery with mass

$$M = \frac{2 \times 10^9\,\text{J}}{6 \times 10^5\,\text{J/kg}} = 3 \times 10^3\,\text{kg}$$

or three *tons*.

Now, this is not quite fair, even by the imprecise standards of this book. When we convert chemical energy to mechanical energy by burning fuel in an engine or a generator, the efficiency is only about one-third. About two-thirds of the chemical energy is lost to heat. Since we have already generated the electrical energy (typically by burning fossil fuel), we need only one-third as much electrical energy as gasoline energy. Therefore, we need only(!) one ton of batteries.

Note that this still does not give you a heating system in your car. Car heaters use some of the two-thirds of the fuel's energy that is converted to heat in order to warm the passenger compartment. Battery-powered electric cars will need to have special air conditioners (called heat pumps) that can be run backward to heat the passenger compartment (thereby cooling the outside).

Since car weights range from one ton (compact car) to two tons (minivan) to three tons (Hummer H2), adding one ton of batteries is significant.

This battery weight is not surprising. Battery technology currently limits many consumer products, from laptop computers to cell phones to cordless power tools. We need to increase the energy density of batteries (with no degradation of lifetime or safety or environmental effects) by a factor of at least five.*

Well, what are you waiting for? Get to work!

* We'll discuss energy transfer speeds later in this chapter.

7.2 Food Is Energy

We get energy (perhaps too much energy) from the food we eat.* The amount of energy in food is listed on the nutrition statement on the back of the package. Europeans find this conveniently measured in joules; Americans need to convert from calories to joules. One food Calorie (C) equals 10^3 physics calories (c) equals 4×10^3 joules (J).†

---

* This energy comes from chemical reactions of the food molecules with oxygen in our body (oxidation). Rapid oxidation is called burning. The energy content of food is measured by burning the food and measuring the energy released.

† Calories were originally used to measure heat and joules were originally used to measure kinetic and potential energy. It was only later that we learned that they are just different forms of energy. A physics calorie is the amount of energy needed to raise the temperature of 1 g of water 1 °C.

Eat here, get gas

7.2.1

How much energy does a typical well-fed
human consume in one year (in joules)?
How much energy does a
typical well-fed car
consume in one
year (in joules)?

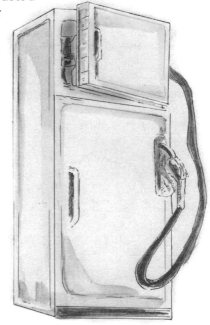

* * * * * * * * * * * * * * * *

HINT: How many food Calories do you consume every day?

HINT: Nutrition guidelines in the US are based on between 2000
and 2500 Calories per day.

HINT: One liter of gasoline contains 3×10^7 J of chemical energy.
1 gal $\approx$ 4 L.

HINT: How much gas does your car use in a year?

HINT: How often do you refill the gas tank?

ANSWER: We consume between 2000 and 3000 food Calories per day. This means that in one year we consume

E = energy per day × days per year

$\quad = 2.5 \times 10^3 \, \text{C/day} \times 4 \times 10^3 \, \text{J/C} \times 4 \times 10^2 \, \text{days/year}$

$\quad = 4 \times 10^9 \, \text{J/year}$

This is an average. Athletes consume more, couch potatoes should consume less.

Now let's consider the hungry automobile. The energy used equals the gasoline used times the energy density of gasoline. From a previous question, we know the energy density of gasoline is 3×10^7 J/L. So how much gas do we use?

In a previous chapter we estimated that the average American drives 10^4 miles each year and gets about 20 miles per gallon. Using these numbers, we can calculate the average car's yearly gasoline consumption

$V = \dfrac{10^4 \, \text{mi/yr}}{20 \, \text{mi/gal}} \times 4 \text{L/gal}$

$\quad = 500 \, \text{gal/yr} \times 4 \, \text{L/gal}$

$\quad = 2 \times 10^3 \, \text{L/yr}$

Thus, the average car uses 500 gallons or 2000 liters per year.

Alternatively, you can estimate your own gasoline consumption by multiplying how often you refuel your car with how many gallons you put in each time. For example, if you refuel your car every week and put in 12 gallons each time, your car's yearly gasoline consumption is

$V = 12 \, \text{gal/refuel} \times 1 \, \text{refuel/week} \times 50 \, \text{weeks/yr}$

$\quad = 600 \, \text{gal/yr} \times 4 \, \text{L/gal} = 2.4 \times 10^3 \, \text{L/yr}$

Gasoline has an energy density of 3×10^7 J/L. Thus, in one year the average car uses energy

$$E = 2 \times 10^3 \text{ L/yr} \times 3 \times 10^7 \text{ J/L}$$

$$= 6 \times 10^{10} \text{ J/yr}$$

That is 60 billion joules. Rather a lot!

Thus, your car consumes about 15 times more energy than you do. This is not surprising, since most humans cannot push one ton along the road at 30 mph.

Farmland for ethanol

7.2.2

How much more farmland would America need to farm to grow corn for enough ethanol to completely replace the gasoline used in all of our cars? Give your answer as a multiple of the amount of farmland we use today. Note that this question assumes, first, that cars can run on 100% ethanol fuel, second, that we use only the human-edible parts of the plant for ethanol, and, third, that we get a lot more energy from burning the ethanol than it takes to produce it.

HINT: What is the ratio of the energy consumed by a car to the energy consumed by its driver? How much more energy does a car consume than a person?

ANSWER: All the energy consumed (i.e., eaten) by humans comes from growing crops or from animals fed by those crops. Almost all of the energy consumed (i.e., burned) by cars comes from fossil fuels. If we use crops to fuel cars, we will need more farmland.

We can assume that the amount of farmland we need is proportional to the amount of energy we need from crops. The typical American or European consumes 4×10^9 J/yr (4 GJ/yr) of energy. The typical American car consumes 6×10^{10} J/yr (60 GJ/yr), or about 15 times as much.

However, we need to include the effects of eating meat. Much of the American and European diet comes from meat, not from vegetables. The conversion efficiency (the ratio of animal feed to animal weight) varies from two for chickens and fish through four for pork to seven for beef [17]. Thus, on average, every calorie that comes from meat represents four calories that come from grain and other animal feed. If half of our food calories come from meat, then we really consume 2 GJ/yr of plants directly and another $4 \times 2 = 8$ GJ/yr of plants indirectly. Thus, we consume 10 GJ/yr of plant energy.

There is about one car for every two Americans, so the energy consumed by one American plus her share of the car is

$$E = 10\,\text{GJ} + \frac{1}{2} 60\,\text{GJ}$$
$$= 40\,\text{GJ}$$

Thus, instead of depending on farmland to provide 10 GJ/yr for each American, we will now need farmland to provide 40 GJ/yr! Yikes! This is a huge increase.

To feed both cars and people from farmland, we would have to vastly increase the farmland under cultivation with a consequent loss of wilderness and animal habitat.

It's not even clear how much gasoline we would save by doing this. That depends on how much fossil fuel it takes to produce one gallon of ethanol from corn. Experts are still wrangling about this. Some even claim that it takes more fossil fuel energy to produce the ethanol than the ethanol contains.

7.3 Power!

In physics, "power" is just the rate at which we use energy. We measure power in watts (W). 1 watt = 1 joule/second. Your 100-W lightbulb uses 100 joules of energy every second. Thus, in one year, your 100-W bulb could use as much as

$$E = 100\,\text{W} \times \pi \times 10^7\,\text{s/yr} = 3 \times 10^9\,\text{J/yr}$$

Wow! Your parents were right when they told you to turn off the lights.

Rather than measuring energy in joules, power companies persist in using kilowatt-hours which is the energy consumed when you use 1 kW for 1 hour. This is

$$1\,\text{kW-hr} = 1000\,\text{W} \times 1\,\text{hr} \times \frac{60\,\text{min}}{1\,\text{hr}} \times \frac{60\,\text{s}}{1\,\text{min}}$$

$$= 3.6 \times 10^6\,\text{J}$$

Thus, your 100-W light bulb uses about 10^3 kW-hr per year.

Hot humans

What is the power (in heat) output
of a human (in W or J/s)?

ANSWER: To answer this problem, we need to estimate our energy consumption over some reasonable period. This is typically given per day. We consume about 2500 food calories per day. Since this is already in units of power (i.e., energy/time) we just need to convert the units:

$$P = 2.5 \times 10^3 \, \text{Cal/day} \times \frac{4 \times 10^3 \, \text{J}}{1 \, \text{Cal}} \times \frac{1 \, \text{day}}{10^5 \, \text{s}}$$

$$= \frac{10^7 \, \text{J}}{10^5 \, \text{s}}$$

$$= 100 \, \text{J/s} = 100 \, \text{W}$$

Alternatively, we could have used the answer to problem 7.2.1 where we estimated that we each consume 4×10^9 J/yr. Converting this to watts, we get

$$P = \frac{4 \times 10^9 \, \text{J/yr}}{\pi \times 10^7 \, \text{s/yr}} = 100 \, \text{W}$$

Thus, humans have the same heat output as a 100-W light bulb. Fifteen of us are equivalent to a 1500-W space heater.

Architects and engineers must take this heat output into account when designing heating and cooling systems for theaters, airplanes, and other structures containing large numbers of people.

Fill 'er up with gasoline

7·3·2

At what rate (in watts) is energy transferred to your car's gas tank when you fill the tank?

HINT: Power = energy/time.

HINT: About how much time does it take to fill the tank? Include only the time you spend pumping the gas.

HINT: Each liter of gasoline contains 3×10^7 J of chemical energy.

HINT: How much gas do you put in the tank?

ANSWER: When you fill your car's gas tank, you are transferring energy. Since power is energy divided by time, we need to estimate the energy transferred by the gasoline and the time it takes to fill the tank. To get the energy transferred, we first need the amount of gasoline transferred when you fill your car's tank. We don't know what car you drive, so we'll use an average car.

A car's gas tank can hold between 10 and 30 gallons. Twenty gallons of gas contains

$$E = 20 \, \text{gal} \times 4 \, \text{L/gal} \times 3 \times 10^7 \, \text{J/L}$$

$$= 2 \times 10^9 \, \text{J}$$

of chemical energy.

Now we need to estimate the time it takes to transfer this energy. We are not counting the time it takes to drive to the gas station or the time it takes to pay for the gas, only the time it takes to pump the gas. It takes more than 1 minute and less than 10 minutes to pump the gas. We'll choose 3 minutes as a reasonable average. (If you need to be more precise, use a stopwatch and time it yourself.)

Thus, the power transfer at a gas station is

$$P = \frac{2 \times 10^9 \, \text{J}}{3 \, \text{min} \times 60 \, \text{s/min}}$$

$$= \frac{2 \times 10^9 \, \text{J}}{2 \times 10^2 \, \text{s}}$$

$$= 10^7 \, \text{W}$$

$$= 10 \, \text{MW}$$

Wow! We transfer chemical energy to our car at a rate of 10 megawatts! That is pretty fast!

Fill 'er up with electricity

At what rate can you transfer electrical
energy to your electric car? Assume
that you are plugging the car
in overnight at home.

HINT: What is the maximum electrical power your house uses?

HINT: What is the maximum electrical power one circuit in your
house can provide?

HINT: What is the power used by a space heater or a microwave
oven?

HINT: Space heaters use about 1.5 kW, about the maximum
power that a typical 20-amp house circuit can deliver.

HINT: How many circuits (i.e., fuses or circuit breakers) do you
have in your home?

ANSWER: You drive your electric car home and plug it in. The energy transfer rate will be limited by the power that your electric circuit can transfer. If you have a special charging station built for the car, then the power will be limited to the power that your house can transfer. Let's consider the power limitations of American houses.

We can find this out in a few ways. We could go around our house and plug in as many space heaters as possible without tripping our circuit breakers. That is too much work and too expensive. A 1.5-kW space heater uses about the maximum possible power for one electrical circuit. One circuit typically supplies several outlets and lights. You can count the number of independent electrical circuits by counting the number of circuit breakers in your electrical panel. Most houses have about ten or twenty. This means that your home has a limit of about

$$P_{max} = 20 \times 1.5\,kW = 30\,kW$$

This will overestimate the limit since electrical wiring is not designed to have all circuits drawing their maximum current simultaneously.

Another way to do this is to look at the main circuit breaker that limits the total electrical current coming into our house. A typical main circuit breaker for a medium-sized house might limit the electrical current to about 125 amps. Since power = voltage × current, this means that the maximum power available is

$$P = V \times I = 110\,V \times 125\,A = 10^4\,W$$

This gives a limit of 10 kW.

Thus, by plugging in an electric car, we can recharge it at a rate of at most 10 kW, about one thousand (!) times slower than refueling it with gasoline. Note that this only works with a special dedicated circuit.

Regular wall outlets would recharge the battery ten times slower still.

This is OK if we are always willing to recharge our electric cars overnight. This would be seriously inconvenient if we needed to refuel our electric car in the middle of a long trip.

Note that this is why automakers are working on developing a plug-in hydrid car. The batteries alone would give the car a 40-mile range, enough for most daily driving. The gas tank and gas engine would be available for the occasional longer trip (or faster acceleration or using the heater or air conditioner).

The Earth, the Moon, and Lots of Gerbils

Now it's time to look at more cosmic questions, concerning meteors, moons, planets, stars, and gerbils.

Chapter 8

✳ ✳ ✳ ✳ ✳ ✳ ✳ ✳ ✳ ✳ ✳ ✳ ✳ ✳ ✳ ✳

"And yet it moves" (e pur si muove)

What is the orbital speed of the Earth around the Sun?
What is its kinetic energy?

* * * * * * * * * * * * * * *

HINT: How much time does it take the Earth to complete one
orbit around the Sun?

HINT: What is the circumference of the Earth's orbit?

HINT: The radius of the Earth's orbit is 1.5×10^{11} m.

HINT: Remember that kinetic energy, the energy of motion, is
proportional to the mass times the square of the speed or
$KE = \frac{1}{2}mv^2$.

HINT: What is the mass of the Earth?

HINT: The radius of the Earth is 6×10^6 m.

HINT: The density of the Earth is more than water (10^3 kg/m^3)
and less than iron (8×10^3 kg/m^3).

ANSWER: To estimate the speed, we need the distance traveled and the time elapsed. The Earth takes one year to travel entirely around the Sun. The distance traveled is the circumference of the circle. The radius of that circle is the distance from the Earth to the Sun, or $R = 1.5 \times 10^{11}$ m. If you used $R = 93$ million miles, that's fine too (93 million miles = 150 million kilometers = 1.5×10^{11} m). Thus, the Earth's speed is

$$v = \frac{\text{distance traveled}}{\text{time}}$$

$$= \frac{2\pi \times 1.5 \times 10^{11} \text{ m}}{1 \text{ year}} = \frac{2\pi \times 1.5 \times 10^{11} \text{ m}}{\pi \times 10^7 \text{ s}}$$

$$= 3 \times 10^4 \text{ m/s}$$

Thus, the Earth has a speed of 60 thousand miles per hour (since $1 \text{ m/s} \approx 2 \text{ mph}$). Rather a lot.

To estimate the kinetic energy, we also need the mass of the Earth. We could Google that, but we would not learn anything from the exercise. Let's estimate it. There are several ways to do this. If we know the formula relating the mass of the Earth and the gravitational acceleration at its surface, we could use that*. Instead we'll estimate the mass from the Earth's volume and its density. We know the radius of the Earth, $R = 6 \times 10^3$ km or 6×10^6 m. If you remember the formula for the volume of a sphere, you can use $V = (4/3)\pi R^3$. If you don't remember that formula, you can use the fact that a sphere is about half the size of a cube with sides equal to the diameter of the sphere ($V < (2R)^3 = 8R^3$). This gives a volume of

$$V = \frac{4}{3}\pi R^3 = 4(6 \times 10^6 \text{ m})^3 = 10^{21} \text{ m}^3$$

---

* The physicists among us can use $g = GM/R^2$, where the gravitational acceleration $g = 10 \text{ m/s}^2$, Newton's constant $G \approx 7 \times 10^{-11}$ N-m^2/kg^2, and R is the Earth's radius.

(where we made the approximation that $\pi = 3$, appropriate for this book and for miscellaneous state legislatures).

We know the Earth is denser than water (10^3 kg/m^3) and less dense than iron (8×10^3 kg/m^3). We'll use a density of 3×10^3 kg/m^3. This gives a mass of

$$M_{Earth} = dV = 3 \times 10^3 \text{ kg/m}^3 \times 10^{21} \text{ m}^3 = 3 \times 10^{24} \text{ kg}$$

The actual mass of the Earth is twice as much, or 6×10^{24} kg. This indicates that the Earth's density is about 6, or much closer to iron than we estimated.

Now we can calculate the kinetic energy of the Earth orbiting the Sun:

$$KE_{Earth} = \frac{1}{2}mv^2 = \frac{1}{2} \times 6 \times 10^{24} \text{ kg} \times (3 \times 10^4 \text{ m/s})^2$$
$$= 3 \times 10^{33} \text{ J}$$

Now *that* is a lot of energy!

Duck!

What is the kinetic energy (in joules
and in megatons of TNT [1 kg of
TNT contains 4×10^6 J of
chemical energy])
of a 1-km meteorite
when it hits the
Earth?

✻ ✻ ✻ ✻ ✻ ✻ ✻ ✻ ✻ ✻ ✻ ✻ ✻ ✻ ✻

HINT: What is the mass of the meteorite?

HINT: Estimate the density of the meteorite.

HINT: What is the speed of the meteorite?

HINT: The meteorite is orbiting the Sun. So is the Earth.

HINT: The meteorite has about the same speed as the Earth in
its orbit.

ANSWER: To estimate the kinetic energy of the meteorite, we need to estimate its mass and its speed. We will estimate the meteorite's density and calculate its mass from its density and volume. Since the problem does not specify whether the meteorite is spherical or cubic or whether the size refers to the radius or the diameter, we will make assumptions that make life easier for us. In this case, we will assume that the meteorite is a cube. This gives a volume $V = (1\,\text{km})^3 = 1\,\text{km}^3$. A sphere of radius 1 km will have a volume four times larger and a sphere of diameter 1 km will have a volume half as much.

Now we need to estimate the density. Again the density is somewhere between those of water and iron. We know that some meteorites are iron and some are rock. We will use the density of iron so the impact is more spectacular. (You can divide that by two if your meteorite is rocky.) This means that the mass of the meteorite is

$$m = Vd = 1\,\text{km}^3 \times \left(\frac{10^3\,\text{m}}{1\,\text{km}}\right)^3 \times 8 \times 10^3\,\text{kg/m}^3$$

$$= 8 \times 10^{12}\,\text{kg}$$

where we needed the factor of $[(10^3\,\text{m})/(1\,\text{km})]^3 = 10^9\,\text{m}^3/\text{km}^3$ to convert from km^3 to m^3. Be very careful with factors like that. If you put them in backward or upside-down, you will make a mistake of 10^{18}. That is a huge error, even in this book.

We now estimate the speed of the meteorite. The speed of the Earth in its orbit is 3×10^4 m/s. This speed comes from the Earth's orbit in the Sun's gravitational field. The meteorite is also orbiting the Sun. Therefore, it will also have a speed reasonably close to the Earth's (i.e., within a factor of two). The meteor could hit the Earth head on (so the relative speed would be 6×10^4 m/s) or it could bump the Earth gently from behind (so the relative speed would be only 1×10^4 m/s or less). We'll just use 3×10^4 m/s for the impact speed.

The kinetic energy of the meteorite as it hits the Earth is thus

$$KE_{\text{meteorite}} = \frac{1}{2}mv^2 = \frac{1}{2} \times 8 \times 10^{12} \text{ kg} \times (3 \times 10^4 \text{ m/s})^2$$
$$= 4 \times 10^{21} \text{ J}$$

This is much less than the kinetic energy of the Earth orbiting the Sun so the meteorite will not change our orbit. However, at 4×10^9 J/ton of TNT, this is the energy contained in 10^{12} tons or 10^6 megatons of TNT. This is rather a lot.

According to [18], a 700-m-diameter meteorite will have a yield of 10^4 to 10^5 megatons and would destroy an area the size of a moderate state (e.g., Virginia). A meteorite this size is expected to hit the Earth every 10^5 years.

The dinosaur killer meteorite is estimated to have been 10 km in size. Small wonder that it caused massive extinctions.

Don't forget to duck.

Super-sized Sun

What is the radius
of the Sun?
What is the Sun's
average density?
Note that the mass
of the Sun is about
one million times
the mass of the Earth,
or 2×10^{30} kg.

* * * * * * * * * * * * * *

HINT: You can cover the image of the Sun with a finger on your
outstretched arm.

HINT: The ratio of the Sun's width to its distance is equal to the
ratio of your finger's width to its distance.

HINT: The distance from the Earth to the Sun is 93 million miles
or 1.5×10^{11} m.

ANSWER: Here we use our knowledge of our own bodies and of the distance to the Sun to determine the radius of the Sun.

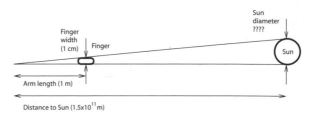

Finger width (1 cm) | Finger

Arm length (1 m)

Sun diameter ????

Sun

Distance to Sun (1.5×10^{11} m)

If we hold a finger out at arm's length, it will cover the image of the Sun. This means that the finger and the Sun both subtend the same angle. This means that the size of the finger divided by the distance to the finger equals the size of the Sun divided by the distance to the Sun. See the figure (which is certainly not to scale). The width of my finger is about 1 cm = 10^{-2} m. The length of my arm is about 1 m. The distance to the Sun is 1.5×10^{11} m. Thus,

$$\frac{\text{Sun size}}{\text{distance to Sun}} = \frac{\text{finger size}}{\text{distance to finger}}$$

$$\text{Sun size} = \frac{\text{finger size}}{\text{distance to finger}} \times \text{distance to Sun}$$

$$= \frac{10^{-2} \text{ m}}{1 \text{ m}} \times 1.5 \times 10^{11} \text{ m}$$

$$= 1.5 \times 10^{9} \text{ m}$$

Note that this is the distance from one side of the Sun to the other, or its diameter. This is actually correct to within 10%! Humans really are the measure of the universe!

Now we can calculate the density of the Sun using $M_{\text{Sun}} = 2 \times 10^{30}$ kg. Rather than looking up the mass of the Sun, you might remember from elementary school that the mass of the Sun is about a million times greater than the mass of the Earth.* The density is the mass divided by the volume:

$$d = \frac{M_{\text{Sun}}}{V} = \frac{M_{\text{Sun}}}{\frac{4}{3}\pi R^3}$$

$$= \frac{2 \times 10^{30} \text{ kg}}{4(7 \times 10^8 \text{ m})^3}$$

$$= \frac{2 \times 10^{30} \text{ kg}}{1.4 \times 10^{27} \text{ m}^3}$$

$$= 1.4 \times 10^3 \text{ kg/m}^3$$

Wow! This is only slightly more than the density of water. The low density of the outer gas envelope must compensate for the very high density of the center of the Sun.

* We could also calculate the mass of the Sun from the period of the Earth's orbit and the gravitational acceleration of the Earth caused by the Sun, but that is beyond the scope of this book

What is the power output of the Sun (in W)?

HINT: There is about 1400 W/m² of solar energy at Earth orbit.
That is 1400 J passing through each square meter every second.

HINT: This power density is uniform over a sphere of radius
$R = 1.5 \times 10^{11}$ m.

ANSWER: The solar power density at Earth orbit is $1400\,\text{W/m}^2$, or $1400\,\text{J}$ passing through each square meter every second. This means that all points that are the same distance from the Sun experience the same solar power density. Since we are in outer space, we do not have to worry about clouds or nighttime diminishing the solar energy. The distance from the Sun to the Earth is $R = 1.5 \times 10^8$ km or 1.5×10^{11} m. Thus, we need to find the total area of all points that are a distance R from the Sun. These points form the surface of a sphere centered on the Sun. The area of this sphere is

$$A = 4\pi R^2 = 12 \times (1.5 \times 10^{11}\,\text{m})^2$$
$$= 2.5 \times 10^{23}\,\text{m}^2$$

The total power output of the Sun is thus the power density times the area:

$$P = \text{solar power density} \times \text{sphere area}$$
$$= 1.4 \times 10^3\,\text{W/m}^2 \times 2.5 \times 10^{23}\,\text{m}^2$$
$$= 4 \times 10^{26}\,\text{W}$$

The actual answer is 3.6×10^{26} W so our answers agree when rounded to one digit.

What relevance does this have to our lives? The total solar energy output determines when we will face the ultimate energy crisis. Forget about high oil prices, forget about running out of oil, forget about oil shale and tar sands. Those are only minor problems. The ultimate limit on human energy consumption will be the total energy output of the Sun.

In one year, the Sun emits energy

$$E = 4 \times 10^{26}\,\text{W} \times \pi \times 10^7\,\text{s/yr} = 10^{34}\,\text{J/yr}$$

In 2003 humans used approximately 4×10^{20} J of energy [19]. This means that we will ultimately be able to increase our energy output by a factor of $10^{34}/4 \times 10^{20} = 2 \times 10^{13}$ (20 trillion). This sounds like a lot. Heck, it *is* a lot.

The US Department of Energy projects that energy use will continue to increase by about 2% per year for the next 25 years. That does not seem like much. Let's see how long it will take us to reach that if our energy use keeps increasing by a mere 2% per year. In one year, our energy usage will increase by a factor of 1.02, in two years by a factor of $(1.02)^2 = 1.04$, in three years by $(1.02)^3 = 1.06$, and in 100 years by $(1.02)^{100} \approx 7$. This is not getting us close. Let's take a short cut. If $1.02^n = 2 \times 10^{13}$, then $n = \log(2 \times 10^{13})/\log(1.02) =$ 1550.* Thus, 1550 years from now (in the year 3550), energy use will have increased by a factor of

$$F = (1.02)^{1550} = 2 \times 10^{13}$$

and we will be using every single watt the Sun puts out!

The *real* energy crisis is coming! Doomsday strikes in 3550!

What this exercise really shows is the absurdity of extrapolating exponential growth trends too far into the future.

---

* Sorry about using logs here. Just remember: A good logarithm enables lumberjacks to be natural musicians.

Gerbils 1, Sun 0

8.5

If the Sun were made out of gerbils, then the Earth would be incinerated. Compare the power output per mass of the Sun and a small mammal.

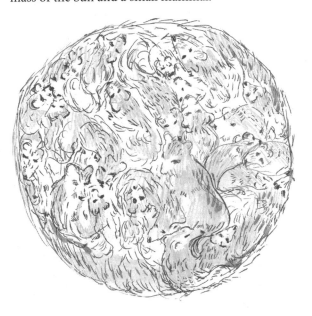

✳ ✳ ✳ ✳ ✳ ✳ ✳ ✳ ✳ ✳ ✳ ✳ ✳ ✳

HINT: Compute the ratio of the Sun's power output to the Sun's mass.

HINT: Compute the ratio of a gerbil's power (i.e., heat) output to a gerbil's mass.

HINT: Since all mammals are warmblooded, the ratio of power output to mass for a gerbil and for a human differ only by about a factor of ten. (The ratio for a gerbil is greater because it has less thermal insulation than a human.)

ANSWER: We can estimate this one of two ways. We can estimate the power output divided by the mass for the Sun and for gerbils or we can estimate the total power output from a mass of gerbils equal to that of the Sun. Let's start with the ratios. The Sun is easy because we have already estimated its power output ($P_{Sun} = 4 \times 10^{26}$ W) and mass ($M_{Sun} = 2 \times 10^{30}$ kg). Thus, the Sun's power density is

$$\mathcal{P}_{Sun} = \frac{P}{M} = \frac{4 \times 10^{26}\,\text{W}}{2 \times 10^{30}\,\text{kg}} = 2 \times 10^{-4}\,\text{W/kg}.$$

The power density of a gerbil is more difficult. We have already estimated the power output of a human. Since both gerbils and humans are mammals, we can expect that the metabolic rates are somewhat similar. However, since humans are much bigger, they have much more thermal insulation and therefore radiate comparatively less heat. The linear scale of a human is about 1 m (it is certainly more than 10 cm [4 in.] and less than 10 m [30 ft]). The linear scale of a gerbil is about 10 cm (it is more than 1 cm and less than 1 m). Thus, humans have about ten times the linear size and thus about ten times thicker insulation. This means that gerbils emit about ten times more heat (per unit mass) than humans.

We will calculate the power density for a human and adjust it upward for a gerbil. A human has a mass of about 100 kg and a power output of about 100 W (see section 7.3.1). This means that the human power density is about

$$\mathcal{P}_{human} = \frac{P}{M} = \frac{100\,\text{W}}{100\,\text{kg}} = 1\,\text{W/kg}$$

The power density of a gerbil will be even greater.

Thus, a gerbil's power density will be at least 10^4 times greater than the Sun's. This means that if the Sun were made of gerbils, it would emit 10^4 times more power. This would increase the surface temperature of the Earth by a factor of ten,* from 300 K (25 °C or 70 °F) to 3000 K (3000 °C or 5000 °F). This would get mighty uncomfortable (although not as uncomfortable as all those poor gerbils).

It is quite unexpected that the Sun, which is powered by nuclear reactions, is less powerful (pound for pound or kilogram for kilogram) than gerbils, which are powered by chemical reactions.

This is, of course, a very silly problem. The Sun carries all of its fuel for billions of years; gerbils do not. If we include all the food, water, and oxygen needed by the gerbils for ten billion years, then the Sun would be way ahead. A human eats about two pounds (1 kg) of food per day. This means that in one year we eat several times our body weight. In a billion years, we would eat several billion times our body weight. This more than makes up for the mere factor of 10^4.

---

* Since power is proportional to temperature to the fourth, $P \propto T^4$.

Chemical Sun

8.6

If the Sun were powered only by
chemical reactions, for how
long could it continue to
burn at its current power
output?

* * * * * * * * * * * * *

HINT: The power output of the Sun is 4×10^{26} W.

HINT: The mass of the Sun is $M_{Sun} = 2 \times 10^{30}$ kg.

HINT: If the Sun were made of gasoline, how much chemical
energy would it contain?

HINT: Ignore the oxygen. It only changes the result by a factor
of three.

HINT: Gasoline contains about 4×10^7 J/kg.

ANSWER: The Sun can contain at most $M = 2 \times 10^{30}$ kg of chemical fuel. We can estimate the maximum energy contained either by assuming that the Sun is made out of gasoline (which we have already calculated) and oxygen or by investigating the optimal fuel. Let's start with gasoline because it is easier. The energy density of gasoline (CH_2 in this book) is about 4×10^7 J/kg. If we ignore the oxygen for now, this gives a total energy content of the Sun

$$E_{Sun} = \text{mass} \times \text{energy density}$$

$$= 2 \times 10^{30} \text{ kg} \times 4 \times 10^7 \text{ J/kg}$$

$$= 8 \times 10^{37} \text{ J}$$

Since the Sun emits energy at a rate of $P = 4 \times 10^{26}$ J/s, this means that this amount of energy will last a time

$$T_{Sun} = \frac{\text{energy content}}{\text{power output}} = \frac{8 \times 10^{37} \text{ J}}{4 \times 10^{26} \text{ J/s}}$$

$$= 2 \times 10^{11} \text{ s} \times \frac{1 \text{ year}}{\pi \times 10^7 \text{ s}}$$

$$= 10^4 \text{ years}$$

That is not a very long time. Human agriculture started about 10^4 years ago. The time is even shorter if we replace about 2/3 of the gasoline with the oxygen needed to burn the gasoline.

We can probably get more energy by oxidizing either pure hydrogen or some other fuel. However, since we cannot do a factor of ten better, it is not worth the effort to calculate it.

This fact bothered 19th century physicists tremendously. William Thomson, who was raised to the British peerage as Lord Kelvin for his great discoveries in thermodynamics, wrote about this in 1862 [20]. He found that the most energetic chemical reaction known could power the Sun for only 3000 years. He further calculated that gravitational potential energy, the energy generated by objects falling into the Sun, could have provided energy for at most 10^7 years. This calculation assumed that the Sun itself was assembled from smaller objects falling in.

This conclusion contradicted the then-recent geological discoveries of the ages of rocks. Kelvin therefore concluded his article thus:

> As for the future, we may say, with equal certainty, that inhabitants of the earth can not continue to enjoy the light and heat essential to their life for many million years longer unless sources now unknown to us are prepared in the great storehouse of creation.

The answer to Kelvin's conundrum is, of course, nuclear energy. We will discuss that in the next chapter.

Nearby supernova

If a star 30 light-years distant went supernova
(and distributed most of its mass in all directions
uniformly), how much of its mass (in kg) would hit
the Earth? A light-year is the distance light travels in
one year so that $1\,\text{l-y} = 3 \times 10^8\,\text{m/s} \times \pi \times 10^7\,\text{s} = 10^{16}$ m. If the planet Krypton was orbiting the star
when it went supernova, how much of its mass (in kg)
would hit the Earth?

* * * * * * * * * * * * * *

HINT: The mass of the Earth is 6×10^{24} kg.

HINT: Replace the mass of the star with the mass of a planet in
the previous part of the question.

HINT: The mass of the Sun is 2×10^{30} kg.

HINT: Stars that go supernova are about ten times more
massive than the Sun.

HINT: Compare the area of the Earth with the area of that
30-light-year sphere.

HINT: The mass of the star would distribute itself uniformly over
the surface of an expanding sphere. As the mass passed our
solar system, that sphere would have a radius of 30 light-years.

ANSWER: Stars get their energy from nuclear fusion. They start by fusing hydrogen into helium. When they run out of hydrogen, they fuse helium into heavier nuclei. Eventually, if the star is large enough ($M_{star} > 8 M_{Sun}$), it will fuse silicon into iron. Once the core of the star becomes iron, fusion no longer produces energy (since iron is about the most tightly bound nucleus). The core then cools and collapses inward. As it collapses, it converts gravitational potential energy to kinetic energy (just like dropping a water balloon from a tall building, only MUCH more so). Some of the energy generated by this collapse creates a giant explosion that blows the outer part of the star rapidly and violently outward.* The remnant of the core then forms either a neutron star or a black hole (depending on the initial mass of the star).

As time passes, the mass of the ejected outer part of the star expands outward in a spherical shell. At some point, part of this shell will pass through our solar system. When it does, the radius of the shell will be 30 light-years. One light-year is the distance light travels in one year: $1 \, \text{l-y} = vt = 3 \times 10^8 \, \text{m/s} \times \pi \times 10^7 \, \text{s} = 10^{16} \, \text{m}$. Therefore, the surface area of the shell at that distance will be

$$A_{shell} = 4\pi R^2 = 12 \times (30 \, \text{l-y})^2$$
$$= 12 \times (30 \times 10^{16} \, \text{m})^2$$
$$= 10^{36} \, \text{m}^2$$

Since the mass of the ejected material is about ten times the mass of the Sun ($M_{Sun} = 2 \times 10^{30}$ kg), we can

---

* You can try this yourself at home on a smaller scale. Place a tennis ball on top of a basketball and drop them together. (It takes a little coordination to drop them so that the tennis ball is still on top of the basketball when they hit the ground.) The basketball plays the part of the falling core. The tennis ball is ejected upward rapidly.

estimate the density of the ejecta as it passes through the solar system:

$$d = \frac{M}{A} = \frac{2 \times 10^{31}\,\text{kg}}{10^{36}\,\text{m}^2} = 2 \times 10^{-5}\,\text{kg/m}^2$$

Now all we need is the area of the Earth in order to figure out how much mass hits the Earth. We could use the surface area ($4\pi R^2$) but none of the mass will hit the far side. To be geometrically correct, we should use the cross-sectional area (πR^2). Using the wrong area will change your answer by only a factor of four. $A_{\text{Earth}} = \pi R^2 = \pi (6 \times 10^6\,\text{m})^2 = 10^{14}\,\text{m}^2$. Thus, the total mass of the supernova ejecta that hits the Earth would be

$$m = \text{area} \times \text{density} = 10^{14}\,\text{m}^2 \times 2 \times 10^{-5}\,\text{kg/m}^2$$

$$= 2 \times 10^9\,\text{kg}$$

That is a lot, but there is only 20 micrograms (a few cubic millimeters of dust) spread over each square meter of surface.

If we now consider how much mass from the planet Krypton would hit the Earth, we need to replace the mass of the star with the mass of the planet. The mass of the Earth is about 10^6 times less than the Sun so we will assume that the mass of Krypton is 10^6 times less than its star. This means that if 2×10^9 kg of star dust hits the Earth, then $10^{-6} \times 2 \times 10^9$ kg $= 2 \times 10^3$ kg of kryptonite reaches the Earth. That is two tons of kryptonite.

Watch out, Superman!

Melting ice caps

How much would the
ocean surface rise
if the ice caps
melted?

HINT: When ice caps on land melt, they raise the water level. When floating ice (e.g., the north polar ice cap) melts, it does not raise the water level because it is already displacing its weight of water.

HINT: Greenland only looks big on a Mercator projection. Ignore it.

HINT: What fraction of the Earth's surface area does Antarctica cover? How many copies of Antarctica would be needed to cover the globe? 1, 10, 100, ...???

HINT: Antarctica lies entirely within the Antarctic circle at 66° south latitude.

HINT: How thick is the ice cap?

HINT: The ice cap is about 2 km thick.

HINT: 1 kg of water has about the same volume as 1 kg of ice.

ANSWER: To solve this problem, we need to estimate the volume of ice resting on the land and then figure out how high it would be if we melted the ice and spread it over the ocean. To do this, we can either estimate the volume of the ice cap and divide by the surface area of the Earth (since the Earth is 3/4 ocean), or we can estimate the height of the ice cap and scale that by the relative areas of the ice cap and the earth.

Since only the land-based ice cap matters, we will ignore the Arctic ice cap. Greenland is much smaller than Antarctica, so we will consider only Antarctica. Let's first estimate the average height of the ice cap. It must be much more than 100 m, since scientists have drilled several-kilometer ice cores that go back 700,000 years. It must be much less than 10^4 m (10 km) since that is the height of Mt. Everest. We'll take the average height to be 10^3 m, the geometric mean of 10^2 and 10^4 m.

Now we need the area of Antarctica. Actually, we just need the relative area of Antarctica and the Earth. If the area of Antarctica is one-tenth the area of the Earth, then the 10^3 m tall ice cap will raise the ocean level by $\frac{1}{10} \times 10^3$ m $= 100$ m. If you look at a globe, you can see that Antarctica is contained well within the Antarctic Circle at 66° south latitude. We'll assume that Antarctica fills the entire circle beyond 70° S. Let's see how many of these circles will go around the world from south to north and back again. Antarctica itself goes from 70° S, to the South Pole at 90° S, and then to 70° S, so that it spans 40° of latitude. The next copy will go from 70° S to 30° S. The third will go from 30° S to 10° N. You get the idea.

There are 360° of latitude as you go around the globe from the South Pole to the North Pole and back again. Therefore, you can place $\frac{360°}{40°} = 9$ copies of Antarctica on that north–south circle. You can place another 9 copies of Antarctica along an east–west circle at the equator. If the surface of the Earth was a square, you could place $9 \times 9 = 80$ copies of Antarctica on it. The actual number will be a bit less than that. We'll estimate that the surface area of the Earth is 50 times greater than the surface area of Antarctica.

Alternatively, you could estimate the area of Antarctica. Assume that Antarctica is a circle. The radius of Antarctica is the distance from the South Pole to $70°$ S, which is $\frac{20°}{360°}$ of the circumference of the Earth or $R_{\text{ant}} = \frac{20°}{360°} \times 2\pi R_{\text{Earth}} = 2 \times 10^3$ km. This gives an area $A_{\text{ant}} = \pi R_{\text{ant}}^2 = 10^7$ km^2. If you compare this to the surface area of the Earth, you will also get a factor of 50.

Now we will take the ice covering each square meter of Antarctica, melt it, and spread it out so that it covers 50 square meters of water. If there are 10^3 m of ice covering Antarctica, then the extra water will raise the ocean height by

$$h = \frac{10^3 \text{ m}}{50} = 20 \text{ m}$$

This would be, shall we say, rather unfortunate for all coastal cities.

The actual expected sea-level rise would be about 80 m [21]. Note that the consequences of an 80-m rise are not that much worse than a 20-m rise. Either would be catastrophic.

To quote Bill Cosby, "How long can you tread water?"

Energy and the Environment

One measure of the advance of civilization is the amount by which we use external sources of energy to multiply our own puny efforts. We have advanced from animal power, through water and wind power, to electrical power generated mostly by fossil fuel.

Chapter 9

✳ ✳ ✳ ✳ ✳ ✳ ✳ ✳ ✳ ✳ ✳ ✳ ✳ ✳ ✳

How much electrical power does one American
or European family
(or household) use
(on average)?

HINT: Remember that power is a measure of how quickly you use energy. Power (in watts) is the amount of energy used (in joules) per second. A 100-W light bulb uses 100 J every second. A kilowatt-hour (kW-hr) is the amount of energy used by ten 100-W light bulbs in one hour or one 100-W light bulb in ten hours.

HINT: Add up the power used by different appliances and light bulbs, including the fraction of the day each one is on.

HINT: Alternatively, take your typical monthly energy bill and assume you paid $0.10 per kW-hr.

ANSWER: There are two ways to estimate the average power we use. We can work from the bottom up by adding up the contributions from individual appliances or we can work from the top down, using our electric bill to estimate the total electric energy we use in a month.

Let's start from the bottom up. Consider the appliances that are on for a significant fraction of the day. Since we are writing this book in August, we will consider air conditioners, stoves and ovens, refrigerators, and light bulbs. We will assume that microwave ovens, water heaters, washing machines, etc. are not on enough to make a big difference.

There are probably five 100-W light bulbs on from 5:00 to 11:00 PM every day. The refrigerator certainly uses more power than a light bulb (100 W) and less than a space- heater (1500 W) so we'll take the geometric mean of 400 W. You can listen to your refrigerator and hear it switch on and off. It is on about one-fourth of the time (that is the time that the compressor is running and it is actively cooling food). The stove and oven probably use more power than a space heater (3×10^3 W) for about an hour a day. The air conditioner is harder to estimate. A room air conditioner can be plugged into a room outlet and therefore uses about the same power as a space heater (1500 W). If you have six rooms, then your total air conditioning (AC) uses about 10 kW. Central air conditioning should be more efficient than individual room air conditioners and probably uses about half that power. Now we want the average AC usage from April to October.* At 2:00 PM

---

* We are assuming a location in about the middle of the US or Europe, where the heating season runs from November to March and the cooling season runs from April to October. We are also assuming that the heating and cooling costs are equal. Your results will vary depend on where you live, but they should be within a factor of ten.

on the hottest day of the year, your AC is probably running constantly. At night in October, your AC is probably unused. During those six months, the AC certainly runs less than 100% and more than 1% of the time. We will take the geometric mean and estimate 10% (or 2.4 hours per day). Let's see what we have so far:

Item	Peak power (kW)	Time used (hr/day)	Average power (kW)
Lights	0.5	6	0.1
Refrigerator	0.4	12	0.2
Stove	3.0	1	0.1
Central AC	5.0	2.4	0.5
Total			0.9

The average household uses an average 0.9 kW of electrical power.

Let's attack the problem from the other end. A typical monthly electric power bill (it's really an energy bill, since we pay for kilowatt-hours, not for kilowatts) is about $100. At an average price of $0.10 per kW-hr, that means we use $P = \$100/(0.10\,\$/\text{kW-hr}) = 10^3$ kW-hr each month. Since there are 24 hours/day × 30 days/month = 700 hours per month, this means that our average power consumption is

$$P = \frac{\text{energy used (in kW-hr)}}{\text{time (in hr)}}$$

$$= \frac{10^3 \text{ kW-hr}}{700 \text{ hr}} = 1.4 \text{ kW}$$

The two methods give the same answer to within a factor of two!

How much electrical power does the US (or Europe)
use? How much electrical energy in one year?

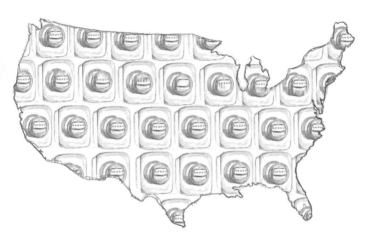

✳ ✳ ✳ ✳ ✳ ✳ ✳ ✳ ✳ ✳ ✳ ✳ ✳ ✳ ✳

HINT: There are $\pi \times 10^7$ s in one year.

HINT: How much extra do you need to add for commercial and
industrial use?

HINT: How many families (or households) are there in the US?

ANSWER: There are two ways to estimate this. We can estimate either the power produced or the power consumed. To estimate the total electrical power consumed by the US or Europe,* we need to add up the total power used by residences, commerce, and industry. We just estimated that one American household uses 1 kW of electrical power. Therefore, we need to estimate the total number of households.

There are 3×10^8 Americans. With two or three Americans per household, there are about 10^8 households. The total residential power usage is then

$$P_{res} = 10^3 \text{ W/household} \times 10^8 \text{ households} = 10^{11} \text{ W}$$

There is probably about as much commercial store and office space as there is residential space, so commercial and residential power use is probably about the same or 10^{11} W.

Industrial power use is probably between one and ten times as much as residential or commercial use, so we will take the geometric mean and estimate a factor of three or 3×10^{11} W.

Thus, the total electrical power used in the US is

$$P_{elec} = 10^{11} \text{ W} + 10^{11} \text{ W} + 3 \times 10^{11} \text{ W} = 5 \times 10^{11} \text{ W}$$

Now let's try to estimate the total power produced. We read somewhere that there are about 100 nuclear power plants and that they produce about 10% of the nation's electricity. Each power plant produces about 1 GW (10^9 W). Thus, the total electrical power produced is about ten times the nuclear power produced or

$$P = 10 \times \text{number of nuclear plants} \times \text{power per plant}$$

$$= 10 \times 10^2 \text{ plants} \times 10^9 \text{ W/plant}$$

$$= 10^{12} \text{ W}$$

---

* The US and Europe have about the same population and standard of living and so consume about the same amount of electricity (at least at the level of precision of this book).

Hmmmm. The two estimates differ by only a factor of two.

Now let's calculate the total electrical energy used by the US in one year. Since energy equals power times time, we have

$$E = P \times t = 10^{12} \, W \times \pi \times 10^7 \, s/yr = 3 \times 10^{19} \, J$$

That is certainly a lot of energy (although it will not get a rocketship to Alpha Centauri).

Now let's compare it to reality. According to the CIA *World Factbook* [22], the US used 3.6×10^{12} kW-hr of electricity in 2003. We need to convert this from kW-hr to J.

$$E = 3.6 \times 10^{12} \, kW\text{-}hr \times \frac{10^3 \, W}{1 \, kW} \times \frac{60 \, s}{1 \, min} \times \frac{60 \, min}{1 \, hr}$$

$$= 1.3 \times 10^{19} \, J$$

We're off by a factor of two.

Solar energy

9·3

How much solar energy reaches the
Earth in a one year?

* * * * * * * * * * * * * * *

HINT: The radius of the Earth is $R \approx 6 \times 10^6$ m.

HINT: There are $\pi \times 10^7$ s/yr.

HINT: Use the cross-sectional area of the Earth, not the
surface area.

HINT: The solar power density at Earth orbit is 1400 W/m².

ANSWER: Since we know that the solar power density is 1400 W/m^2, we need to estimate the time and the area to get the energy received. We cannot just use the surface area of the Earth ($A = 4\pi R^2$) for two reasons: (1) Half of it is dark (i.e., night) and (2) the light intensity decreases at higher latitudes due to the angle of incidence of the light. If we imagine a slice through the middle of the Earth, that circle will be perpendicular to the Sun and all the sunlight will hit it from directly overhead. We will use the area of this circle (the same one we used for the supernova debris). Thus, the area we want is

$$A = \pi R^2 = \pi \times (6 \times 10^6 \text{m})^2 = 10^{14} \text{m}^2$$

Now we can calculate the energy

$$E = 1.4 \times 10^3 \text{ W/m}^2 \times 10^{14} \text{ m}^2 \times \pi \times 10^7 \text{ s/yr}$$
$$= 4 \times 10^{24} \text{ J/yr}$$

This is more energy than we could gain by flattening the Rockies or the Alps.

As we discussed in the answer to question 8.4, humans used 4×10^{20} J of energy in 2003. Thus, we used a fraction

$$f = \frac{4 \times 10^{20} \text{ J/yr}}{4 \times 10^{24} \text{ J/yr}} = 10^{-4}$$

of the available solar energy hitting the Earth. Of course, most of the energy we used came from ancient solar energy as stored in fossil fuels.

This means that we could expand our energy use by a factor of 100 and still only use 1% of the available solar energy. However, at a 2% annual increase in energy use, we will reach that point in a mere 230 years.

How much land (in km^2) would be needed to supply the US electrical energy needs with solar energy? What fraction of the US land area would be needed?

※ ※ ※ ※ ※ ※ ※ ※ ※ ※ ※ ※ ※

HINT: The solar power density above the atmosphere at Earth orbit is 1400 W/m^2.

HINT: Don't forget to include clouds and night.

HINT: What is the efficiency of solar panels for converting solar energy to electrical energy?

HINT: What is the area of the US? Include only the lower 48 states.

HINT: We did this before. See question 3.9.

ANSWER: We will start with the total solar power density outside the atmosphere and then account for the effects of the atmosphere, clouds, night, and conversion efficiency.

The solar constant is 1.4×10^3 W/m². About half of that reaches the ground (at noon on a clear day). Including the effects of clouds and night decreases the total flux by about a factor of ten to 140 W/m². The efficiency of solar panels is about 10%, so they will generate an average of about 14 W/m² of electrical power.

From problem 9.2 we know that the US uses an average of 5×10^{11} W. Thus, we would need an area

$$A_{\text{solar cell}} = \frac{\text{electrical power used}}{\text{electrical power per area}}$$

$$= \frac{5 \times 10^{11} \text{ W}}{1.4 \times 10^1 \text{ W/m}^2}$$

$$= 4 \times 10^{10} \text{ m}^2$$

Now $(1 \text{ km})^2 = (10^3 \text{ m})^2 = 10^6 \text{ m}^2$. Thus, the area needed is $A = 4 \times 10^4 \text{ km}^2$. This is a square that is 200 km (150 mi) on a side.

This seems like a lot of land, but the US is a very big country. Let's compare this area to the area of the US. We already estimated the area of the contiguous US back in question 3.9 as $A = 10^7 \text{ km}^2$.

Therefore, the fraction of US land area needed if we used solar energy cells (photovoltaics) to provide all of our electrical energy would be

$$f = \frac{\text{solar cell area}}{\text{US land area}} = \frac{4 \times 10^4 \text{ km}^2}{10^7 \text{ km}^2} = 4 \times 10^{-3}$$

We would need to use 0.4% of all US land area for solar arrays.

This would be amazingly expensive for three reasons. First, solar cells are very expensive. They cost about \$10 per watt installed (in 2006), so 5×10^{11} W would cost 5 trillion dollars. Second, the best sites for solar power (i.e., deserts) are frequently far from population centers and would require expensive long-distance power transmission. Third, we would need an amazing number of batteries to charge during the day and provide power at night.

How much electric power can a wind turbine (a modern windmill) generate? (Consider the kinetic energy of the air passing through the area swept by the turbine blades.)

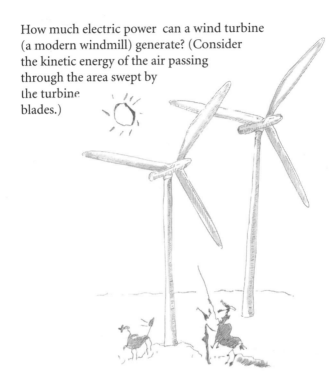

✳ ✳ ✳ ✳ ✳ ✳ ✳ ✳ ✳ ✳ ✳ ✳ ✳ ✳ ✳ ✳

HINT: Modern wind turbines are as tall as a ten-story building.

HINT: In one second, how much air (in kg) passes through the area swept by the blades?

HINT: Pick a reasonable sustained wind speed. Remember that 2 mph ≈ 1 m/s.

HINT: Air has a density of about 1 kg/m³.

HINT: How much kinetic energy does that air have?

HINT: What fraction of the wind energy will the wind turbine convert to electrical energy?

ANSWER: Wind turbines convert wind kinetic energy into electrical energy. Therefore, to estimate the power that a wind turbine can generate, we first need to estimate the available kinetic energy. To estimate that, we need the air mass and the wind speed. The air mass passing through the area swept by the blades will be determined by the area swept by the blades and by the wind speed. We will first estimate the wind speed.

Wind turbines are placed in windy locations. Typical sustained wind speeds will be between 20 and 30 mph or between 10 and 15 m/s. (Less than that would not be considered windy, more than that is rather unlikely.) We'll use 10 m/s because it is a nice round number. We'll see how the final answer depends on wind speed later.

Modern wind turbines are as tall as a ten-story building. (Some are even taller.) At about 4 m or 12 ft per story, the wind turbine would be 40 m tall. This means that the length of the blades is 40 m and the area of the circle swept by the blades is $A = \pi r^2 = 3 \times (40\,\text{m})^2 = 5 \times 10^3\,\text{m}^2$.

At a wind speed of 10 m/s, air travels 10 m in one second. Thus, in one second, a volume of air $V = 10\,\text{m} \times 5 \times 10^3\,\text{m}^2 = 5 \times 10^4\,\text{m}^3$ passes through the area swept by the turbine blades. At a density of $1\,\text{kg/m}^3$, this air has a mass $m = 5 \times 10^4\,\text{kg}$. Thus, 50 tons of air passes through the blade area every second!

Now we can calculate the kinetic energy of one second's amount of air:

$$\text{KE}_{\text{air}} = \frac{1}{2}mv^2 = \frac{1}{2} \times 5 \times 10^4\,\text{kg} \times (10\,\text{m/s})^2$$
$$= 3 \times 10^6\,\text{J}$$

Thus, the power of the wind passing through the turbine blade area is 3×10^6 W or 3 MW. That's a lot of available power.

Now we need to estimate the efficiency of the wind turbine in extracting the kinetic energy from the air and converting it to electrical energy. That efficiency has to be more than 1% (since that is a very low efficiency) and less than 100% (since that would bring the wind to a complete stop). We will take the geometric mean and estimate an efficiency of 10%. This means that the available power from a wind turbine at a constant wind speed of 20 mph (10 m/s) would be 10% of 3×10^6 W, which equals 3×10^5 W or 300 kW.

Now let's consider the effects of wind speed. If we double the wind speed then we double the speed in the kinetic energy equation, which quadruples the kinetic energy (since the speed is squared). However, it also doubles the mass passing through the turbine blade area, so doubling the wind speed increases the available kinetic energy by a factor of $2^3 = 8$. Thus, our 40-m turbine would produce 300 kW at a wind speed of 10 m/s and eight times more, or 2.4 MW, at a wind speed of 20 m/s (40 mph).

Now its time for the dreaded reality check. According to the Danish Wind Industry Association [23], a modern 92-m blade diameter wind turbine with a maximum tolerable wind speed of 25 m/s would have a nominal output of 2.75 MW and an average power of almost 1 MW. Since our hypothetical turbine would produce 2.4 MW at 20 m/s, our estimate is rather good.

Unfortunately, wind energy is not reliable enough to provide base electrical power. Yet another reason for better batteries.

The power of coal

9.6

How much fuel does a 1 GW coal-fired electrical power plant require? Express your answer in kilograms per year and in 100-ton railroad cars per day. Note that the plant uses 3 GW of thermal power to produce 1 GW of electrical power.

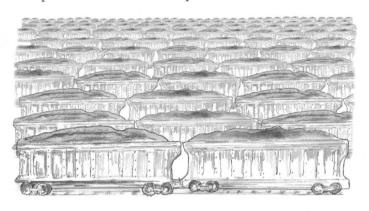

✳ ✳ ✳ ✳ ✳ ✳ ✳ ✳ ✳ ✳ ✳ ✳ ✳ ✳ ✳

HINT: We can get 1.5 eV from burning one atom of Carbon to form CO_2. See section 7.1 for more details.

HINT: One mole of Carbon has a mass of 12 g.

HINT: How many moles of Carbon are in 1 kg of coal?

HINT: Coal is approximately pure carbon.

HINT: How much thermal energy does the power plant use in one year?

ANSWER: To figure out how much coal a 1-GW electric power plant needs per year, we need to estimate the thermal energy needed in one year and the energy density of coal. Oil, gas, coal, and nuclear power plants use the heat from their fuel to boil water and then use the steam to turn a turbine and produce electricity. Only about one-third of the thermal energy is converted to electrical energy; the other two-thirds is emitted as heat. Thus, in one year, a 1-GW electrical power plant will need to burn enough fuel to produce

$$E_{year} = 1 \, GW \times 3 \times \pi \times 10^7 s/yr = 10^{17} \, J/yr$$

This thermal energy comes from the chemical energy of the coal. Burning one atom of carbon will produce 1.5 eV of energy. One mole of carbon has a mass of 12 g = 1.2×10^{-2} kg. Thus there are $1/1.2 \times 10^{-2}$ = 80 moles of carbon in a kilogram. There are 6×10^{23} atoms in a mole. Thus, one kilogram of coal contains chemical energy equal to

$$E_{coal} = \frac{1.5 \, eV}{atom} \times \frac{6 \times 10^{23} \, atoms}{mole}$$
$$\times \frac{2 \times 10^{-19} \, J}{eV} \times \frac{80 \, moles}{kg}$$
$$= 10^7 \, J/kg$$

Alternatively, we could have taken the results for gasoline (CH_2) and divided by two since we get about as much energy from burning the H_2 as from the C and the H_2 has much less mass than the C. This would give an energy density of $\frac{1}{2} \times 4.5 \times 10^7$ J/kg = 2×10^7 J/kg.

When we compare with reality, we find that coal has an energy density of between 10 and 30 MJ/kg [15], so we are within a factor of three.

Now, we can calculate the coal needed to operate the power plant for a year:

$$M_{coal} = \frac{\text{energy needed}}{\text{coal energy density}} = \frac{10^{17} \text{ J/yr}}{2 \times 10^7 \text{ J/kg}}$$

$$= 5 \times 10^9 \text{ kg/yr}$$

This is 5 million tons of coal. That is a lot of coal.

Let's try to get a feel for how much that is. We cannot really picture a ton of coal, let alone 5 million tons. One railroad car can carry about 100 tons of coal. One very long railroad train will have 100 cars. Thus, to carry 5×10^6 tons of coal, we will need 5×10^4 railroad cars, organized into five hundred 100-car trains. This means that our power plant needs more than one 100-car coal train every single day.

Wow. That really is a lot of coal.

The power of nuclei

9·7

There are two general types of nuclear reactions,
fission and fusion. We use fission reactions on Earth
to produce nuclear energy and nuclear explosions. The
Sun and stars use fusion reactions to produce solar
energy. Fissioning (splitting) a heavy element such as
uranium or plutonium produces about 200 MeV
(mega-electron volts) of energy. Fusing four hydrogen
nuclei (protons) into one helium nucleus produces
28 MeV of energy. Nuclear reactions can convert between 0.1 and 1% of the mass of the atoms to energy.
How much fuel does a 1-GW$_e$ nuclear power plant
require in one year?

* * * * * * * * * * * * * *

HINT: Each fission produces about 200 MeV of energy.

HINT: The fuel is only about 5% ^{235}U. The rest is ^{238}U which is
relatively inert.

HINT: The fissionable isotope of Uranium is ^{235}U which has a
mass of 235 g/mole.

HINT: The nuclear power plant needs 3 GW of thermal power to
produce 1 GW$_e$ of electrical power (just like the coal-fired plant of
the previous question).

ANSWER: The nuclear power plant needs the same thermal energy as a coal plant to produce its 1 GW of electrical energy for a year. Thus, from the previous question, we know it needs $E = 10^{17}$ J/yr.

This thermal energy comes from the fission of ^{235}U. Fissioning one atom of ^{235}U will produce about 200 MeV (2×10^8 eV) of energy. One mole of ^{235}U has a mass of 235 g or about 0.24 kg. This means that there are 4 moles/kg. Thus, ^{235}U has an energy density of

$$E_{^{235}\text{U}} = \frac{2 \times 10^8 \text{ eV}}{\text{atom}} \times \frac{6 \times 10^{23} \text{ atoms}}{1 \text{ mole}}$$

$$\times \frac{2 \times 10^{-19} \text{ J}}{1 \text{ eV}} \times \frac{4 \text{ moles}}{\text{kg}} = 8 \times 10^{13} \text{ J/kg}$$

That is 8 million times more energy density than coal. However, we need to reduce this because only 5% of the fuel is ^{235}U. The other 95% is the relatively inert ^{238}U. This means that the effective energy density is only 5% of 8×10^{13} J/kg or $E_{\text{U}} = 4 \times 10^{12}$ J/kg.

This means that in one year, a 1-GW nuclear power plant will need

$$M_{\text{U}} = \frac{\text{energy needed}}{\text{uranium energy density}} = \frac{10^{17} \text{ J/yr}}{4 \times 10^{12} \text{ J/kg}}$$

$$= 2 \times 10^4 \text{kg/yr}$$

or 20 tons of 5% enriched uranium. Since the density of uranium is about 2×10^4 kg/m^3, this mass of uranium would fit in a cube that is 1 m (3 ft) on each side. Thus, the fuel for a nuclear power plant for an entire year would fit under your dining room table.*

---

* We do not suggest storing the fuel there for two reasons: (1) You would bang your shins on it repeatedly, and (2) if the fuel was concentrated into such a small volume with no control rods, it would go critical, the chain reaction would be out of control, and you would have a highly radioactive meltdown in your dining room. But at least your food would not get cold.

Hard surfaces

9.8

What fraction of the US land
area is impervious
(i.e., roofed or paved)?

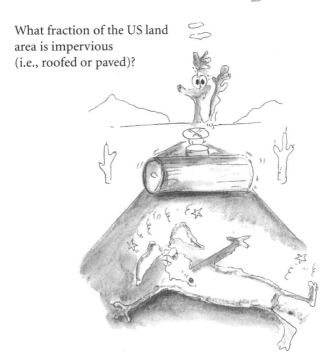

HINT: Most Americans live in single-family homes.

HINT: What is the area of the roof on an average home?

HINT: How do the areas of commercial building compare to
residential ones?

HINT: How much does each road building have?

HINT: There are 3×10^8 Americans.

HINT: What is the area of the US? See question 9.4.

ANSWER: We will estimate the size of the typical American home, estimate the amount of road per home, and add some more space to account for commercial development.

A typical two-story home has about 2000 square feet of living space and thus 1000 square feet of roof area. Since $1\,m^2 = (3\,ft)^2 \approx 10\,ft^2$, the typical home has about $100\,m^2$ of roof area.

Now we need to include roads. Our typical house is about 10 m wide by 10 m deep. The distance between the front doors of adjacent houses will certainly be greater than 10 m (i.e., no space between houses) and less than 100 m (i.e., a football field between houses). We'll take the geometric mean of 10 and 100 m and use a house spacing of 30 m (100 ft). Therefore, each house will have 30 m of road.

The width of this road is shared between houses on both sides of the street. A typical suburban road is more than 1 car (3 m) wide and less than 10 cars (30 m) wide so we will use a width of 10 m (30 ft). This means that the typical house will have a 30-m-long stretch of road that is 5 m wide with an area of $150\,m^2$. Thus, each house will have $100\,m^2$ of roof and $150\,m^2$ of road for a total of $250\,m^2$.

We estimated earlier that the 3×10^8 Americans form about 10^8 households. Thus, residential roofs and roads will cover an area $A = 10^8 \times 2.5 \times 10^2\,m^2 = 2.5 \times 10^{10}\,m^2$.

Stores, offices, and factories will certainly occupy more than one-tenth and less than ten times the amount of residential space so we will estimate that commercial roofs and roads contribute another $2.5 \times 10^{10}\,m^2$. Therefore, the total roofed and paved area in the US is $A = 5 \times 10^{10}\,m^2$ or $5 \times 10^4\,km^2$.

Alternatively, we can estimate this by considering (1) the population density of our cities and (2) the roof and road density of our cities. A typical US city has a density of 2000 people per square mile or 1000 people/km^2. At this density, if everyone lived in cities, the cities would cover an area

$$A = \frac{3 \times 10^8 \text{ people}}{10^3 \text{ people/km}^2} = 3 \times 10^5 \text{ km}^2$$

You can look at Google maps for any suburban region and estimate that this city is about 25% paved and roofed (more than 10% and less than 50%). Therefore, the total paved area is about 10^5 km^2. This estimate agrees (within a factor of two) with the other method. In reality, according to a 2004 study done by the US National Atmospheric and Oceanographic Administration (NOAA) [24], 1.1×10^5 km^2 are impervious. This is slightly higher than either of our estimates.

We already estimated the land area of the US in question 9.4 to be 8×10^6 km^2. Therefore, the fraction of the US that is paved or roofed is

$$F = \frac{\text{roofed and paved area}}{\text{US land area}}$$

$$= \frac{8 \times 10^4 \text{ km}^2}{8 \times 10^6 \text{ km}^2} = 10^{-2}$$

or about 1%. This is about twice the amount of land that we estimated we would need to provide enough solar energy for all of our electrical needs. This puts that number in perspective. To provide enough solar energy for the US, we would have to have an area of solar panels equal to half the area of all the roofs and roads in the entire country. Yikes!

The Atmosphere

We rely on the atmosphere for airplanes, kites, wind-blown hair, and, oh yeah, breathing. Let's find whose breath you just breathed in, how much oxygen plants provide, and how much carbon dioxide cars emit.

Chapter 10

* * * * * * * * * * * * * * * *

Into thin air

10.1

What is the mass of the atmosphere?

HINT: The weight of the atmosphere exerts 15 lb of force on every square inch or 10^5 N on every square meter of the Earth's surface.

HINT: 10^5 N is the weight of 10^4 kg.

HINT: What is the surface area of the Earth?

HINT: The radius of the Earth is 6×10^6 m.

HINT: If you forget the surface area of a sphere, you can just treat the Earth as a cube.

ANSWER: Atmospheric pressure is measured in too many different ways. It is 15 pounds per square inch or 10^5 newtons per square meter or (shudder) 760 mm (30 in.) of mercury or 10 m of water. These last two measures just indicate that the weight of the air over a certain area is the same as the weight of an amount of mercury sufficient to cover the same area to a depth of 760 mm (or enough water to cover the same area to a depth of 10 m).

We can use any of these measures to calculate the mass of the atmosphere. To do that, we just need to multiply the area of the Earth's surface (in appropriate units) by the mass per square whatever. For example, if we use 15 pounds per square inch, we would need to calculate the area of the Earth in square inches (ick!). Alternatively, if we use 760 mm of mercury, we would have to calculate the weight of enough mercury to cover the entire Earth's surface (including the oceans) to a depth of 760 mm (or enough water to cover it to a depth of 10 m).

We will use the pressure of 10^5 N/m^2 so we don't have to calculate square inches. 10^5 N is the weight of 10^4 kg (since $F_{gravity} = mg$, where $g = 10$ m/s^2).

Now we need the surface area of the Earth. We happen to remember that the surface area of a sphere is $A = 4\pi R^2$, but not everyone does. You can also pretend that the Earth is a cube. In that case, the area of each of the six faces is $(2R)^2 = 4R^2$ so that the total area is $A = 24R^2$. This is close enough for this book and saves you from having to memorize pesky formulas. We'll use the correct equation if you don't mind:

$$A_{Earth} = 4\pi R^2 = 12 \times (6 \times 10^6 \text{ m})^2 = 4 \times 10^{14} \text{ m}^2$$

Thus, the mass of all the air is

$$M_{air} = 10^4 \text{ kg/m}^2 \times 4 \times 10^{14} \text{ m}^2 = 4 \times 10^{18} \text{ kg}$$

That is 4 billion billion kilograms or more than the mass of the Alps!

How many molecules
of Alexander the Great's
last breath do you inhale
with each breath?

* * * * * * * * * * * * * * *

HINT: Assume that the atmosphere has mixed thoroughly in the last 2000 years so the molecules are randomly distributed.

HINT: What is the volume of a breath?

HINT: How many molecules are in a breath?

HINT: One mole of gas has 6×10^{23} molecules and occupies 20 L of volume.

HINT: How many breaths are there in the atmosphere?

ANSWER: To answer this question, we need to estimate the number of molecules in Alexander's last breath and the fraction of the atmosphere that you just breathed in a few seconds ago. The second part amounts to estimating the number of breaths in the atmosphere.

The volume of a breath is about 1 L (1 quart). It is certainly more than a cup (1/4 L) and less than a gallon (4 L). We have no idea how many molecules are in a liter, but we do know the number of molecules in a mole. A mole contains 6×10^{23} molecules and takes up 20 L (as a gas at standard temperature and pressure). Thus, a 1-L breath contains 1/20 of a mole and hence contains 3×10^{22} molecules.

Now we need the number of breaths in the atmosphere. We can estimate this using either the volume of a breath (in which case we need the volume of the atmosphere) or the mass of a breath (in which case we can use the mass of the atmosphere that we just estimated). Since we're lazy, we'll use the mass.* The average density of air is about 1000 times less than that of water. Since 1 L of water has a mass of 1 kg, 1 L of air would have a mass of 1 g. You can also calculate the mass of air from its molecular weight. Our 1-L breath is 1/20 of a mole. Air is composed of N_2 (molecular weight $= 2 \times 14 = 28$) and O_2 (molecular weight $= 2 \times 16 = 32$) molecules and so the average molecular weight is 30. One mole of air would have a mass of 30 g. Therefore, air has a density $d = (30\,\text{g/mole})/(20\,\text{L/mole}) = 1.5\,\text{g/L}$. (The difference between 1 g and 1.5 g is due to roundoff error and is negligible for this book.)

---

* This also lets us avoid the question of how high the atmosphere is. Unlike the oceans, the atmosphere thins out as you ascend but never really ends. The boundary of outer space is somewhere between 100 and 300 km (60 to 200 mi). The maximum height at which there is enough air for humans to breathe is a bit less than the height of Mt. Everest (10 km).

The total mass of the atmosphere (from the previous question) is $m_{air} = 4 \times 10^{18}$ kg $= 4 \times 10^{21}$ g. Since one breath has a mass of 1 g, with each breath you inhale one part in 4×10^{21} of the atmosphere. Thus, the number of Alexander the Great's (AG) molecules that you just inhaled is

$N_{AG} =$ AG's molecules in the atmosphere

$\times$ fraction of the atmosphere inhaled

$= 3 \times 10^{22}$ molecules $\times \dfrac{1}{4 \times 10^{21}}$

$= 8$ molecules

Thus, you just inhaled about eight molecules from Alexander the Great's (and Confucius' and Ramses' and . . .) last breath (and second-to-last and . . .).

We hope you treated those molecules with reverence and care.*

---

* Alas, we know from quantum mechanics and thermodynamics that all oxygen molecules are alike and cannot be tagged like classical macroscopic particles [25]. Thus, this question is ultimately meaningless. The following analogy can be helpful. If you pay for a $20 purchase with a $20 bill and then six months later we give you $20 (for being such a loyal reader) then there is a chance that the same bill returned to you (since each has a unique serial number). If, however, you pay for your purchase with a credit card and then six months later we transfer $20 to your bank account, it is completely meaningless to ask whether it is the same bill. However, it is still fascinating to know that are about as many molecules in one breath as there are breaths in the atmosphere.

Suck it up

How much time would it take human respiration to use up 10% of all atmospheric O_2 (oxygen), ignoring all other contributions?

✳ ✳ ✳ ✳ ✳ ✳ ✳ ✳ ✳ ✳ ✳ ✳ ✳ ✳ ✳ ✳ ✳

HINT: How much time would it take for all the people to breathe in (and out) the entire atmosphere?

HINT: There are 6×10^9 people on the Earth.

HINT: How frequently do we breathe?

HINT: What is the mass of one breath?

HINT: Artificial respiration works. There is enough O_2 in the air you exhale to keep someone else alive.

HINT: How much of the inhaled oxygen is used?

HINT: We use oxygen from the air (O_2) to get energy by oxidizing our food. The used oxygen is exhaled as CO_2 or H_2O.

ANSWER: We need to estimate the amount of oxygen used up in each breath. From there, we can go on to estimate the number of breaths it will take to use up 10% of the oxygen in the atmosphere. We know that cardiopulmonary resuscitation (CPR) can keep someone alive until the ambulance arrives. This means that the air we force into someone's lungs by breathing out contains enough oxygen to sustain life (even for someone sick enough to need CPR). This means that we use relatively little of the oxygen in each breath. We estimate that we use 10% of the oxygen that we inhale (a lot more than 1% and a lot less than 100%).

This means that we now need to estimate the amount of time it will take humans to breathe in (and out) the entire atmosphere once (since that will use up 10% of the oxygen). Note that we have completely avoided needing to know the abundance of oxygen in the atmosphere.*

In the previous question we estimated that each breath has a volume of about $1\,\text{L}$ and a mass of about $1\,\text{g}$ and that there are 4×10^{21} breaths in the atmosphere. Therefore, we just need to estimate how much time will be needed for humanity to take those 4×10^{21} breaths. Well, there are 6×10^9 of us. This means that the number of breaths you need to take is

$$N_{\text{breaths}} = \frac{4 \times 10^{21}\ \text{breaths}}{6 \times 10^9} = 7 \times 10^{11}$$

We take about one breath every few seconds. Let's estimate four seconds per breath. This means that it will take each of us a time

$$t = 7 \times 10^{11}\ \text{breaths} \times 4\,\text{s/breath}$$
$$= 3 \times 10^{12}\,\text{s} \times \frac{1\ \text{year}}{\pi \times 10^7\,\text{s}} = 10^5\ \text{years}$$

That is 100 thousand years.

---

* If you're curious, it's about 20%. If you're not curious, why are you reading this footnote?

OK. Individually, we won't survive long enough to accomplish this. However, all the other animals are helping us. Since human (or at least hominid) life on this planet dates back millions of years and animal life on this planet dates back hundreds of millions of years, it's a good thing that plants are continually replenishing the oxygen supply.

CO$_2$ from coal

10.4

How much carbon dioxide (in kg) does a 1-GW$_e$ coal-fired power plant release into the atmosphere each year? What fraction of the mass of the atmosphere is this?

* * * * * * * * * * * * * * *

HINT: See question 9.6.

HINT: How many kilograms of CO$_2$ are produced from each kilogram of C?

HINT: Carbon and oxygen have about the same atomic weight.

ANSWER: The amount of carbon dioxide released into the atmosphere will be directly proportional to the amount of carbon burned by the power plant. If all of the carbon is completely burned, then it will all be emitted from the smokestack in the form of carbon dioxide. Fortunately, we know how much coal the 1-GW$_e$ coal-fired power plant uses in one year. We estimated that in question 9.6 to be 5×10^9 kg. Each carbon atom is oxidized and becomes a molecule of carbon dioxide. Carbon and oxygen are relatively light elements and have about the same atomic weight.* Therefore, each molecule of carbon dioxide (CO_2) has about three times the mass of a carbon atom. Therefore, the total mass of CO_2 emitted into the atmosphere in one year by a 1-GW$_e$ coal-fired power plant is

$$M_{CO_2} = 3 \times 5 \times 10^9 \, \text{kg C} = 2 \times 10^{10} \, \text{kg}$$

That seems to be a lot of CO_2. Let's compare it to the mass of the atmosphere. We calculated the mass of the atmosphere in question 10.1 to be $M_{air} = 4 \times 10^{18}$ kg. This plant will add

$$f = \frac{2 \times 10^{10} \, \text{kg}}{4 \times 10^{18} \, \text{kg}}$$

$$= 5 \times 10^{-9}$$

or five parts per billion CO_2 to the atmosphere every year. If you remember those golf balls circling the equator in the first chapter, that would amount to five CO_2 golf balls out of the billion needed to circle the globe. While this does not seem like much, there are a lot of power plants.

A natural gas-fired power plant, fueled by methane (CH_4), will produce about three times less carbon

---

* If you insist on being precise, the atomic weight of carbon is 12 and the atomic weight of oxygen is 16.

dioxide per GW than a coal-fired plant because it oxidizes the hydrogen and the carbon: the carbon will oxidize to carbon dioxide (CO_2) and the four hydrogen atoms will oxidize to two water molecules (H_2O). With three reactions, it will produce three times as much energy per carbon atom.

Note that carbon dioxide is a colorless gas. The "smoke" you see emerging from a smokestack will be some combination of partially combusted fuel, impurities (e.g., ash), and steam (condensing water vapor).

Note also that carbon dioxide is naturally present in the atmosphere at a level of a few hundred parts per million. Plants breathe in CO_2 and breathe out oxygen. Thus, carbon dioxide is not a standard pollutant (like smog) that harms people directly. Any damage it might do is indirect.

A healthy glow

10.5

How much high-level nuclear (i.e., highly radioactive) waste does a 1-GW$_e$ nuclear power plant produce in a year?

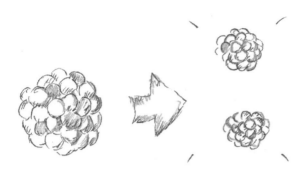

HINT: See question 9.7.

HINT: The used fuel is most of the highly radioactive waste.

ANSWER: The nuclear fuel in the reactor is composed primarily of two isotopes of uranium, ^{235}U and ^{238}U. The ^{235}U fissions into smaller nuclei, many of which are highly radioactive. The daughter nuclei and the ^{238}U in the reactor are exposed to a tremendous amount of radiation, which also produces many highly radioactive by-products. Thus, all of the fuel becomes highly radioactive waste. That is most of the high-level nuclear waste produced by the power plant. Thus, the mass of the highly radioactive waste is between one and ten times the mass of the fuel. We'll take the geometric mean and use a factor of three.

Fortunately, we know how much nuclear fuel the 1-GW$_e$ nuclear power plant uses in one year. We estimated that in question 9.7 to be 2×10^4 kg. Thus, a nuclear power plant produces about

$$M_{\text{nuclear waste}} = 3 \times 2 \times 10^4 \text{ kg} = 6 \times 10^4 \text{ kg}$$

or about 60 tons of highly radioactive nuclear waste per year.

This waste is extremely dangerous. On the other hand, since it so compact (and solid) it should be much easier to dispose of safely.

CO$_2$ from cars

10.6

How much carbon dioxide (in kg)
does one car emit into the
atmosphere each year?
All American cars?

* * * * * * * * * * * * * * * *

HINT: How much gasoline does one car burn in a year?

HINT: How far does each car drive and how many gallons per
mile does it get?

HINT: Each gallon of gas contains 4 L and has a mass of
about 4 kg.

HINT: Each kilogram of carbon in the gasoline combines with
almost 3 more kilograms of oxygen to make CO$_2$.

HINT: There are 3×10^8 Americans.

HINT: How many cars do we each have?

HINT: There is about one car for every two Americans.

ANSWER: We already estimated that the average American car is driven about 10^4 miles per year and gets about 20 miles per gallon. This means that it consumes about $(10^4 \text{ miles per year})/(20 \text{ mpg}) = 500$ gallons per year. At 4 liters per gallon, this is 2000 liters per year.

Now we need to convert from volume to mass. A liter of water has a mass of 1 kg. Gasoline is a bit lighter than water, but not enough for us to worry about. Thus, your car burns about 2000 kg of gasoline per year. That is 2 tons of gasoline, more than the mass of most cars!

Gasoline is a hydrocarbon with about two hydrogen atoms for each carbon atom. Since carbon has an atomic mass of 12 and hydrogen has an atomic mass of 1, we can (and will) ignore the hydrogen. The two oxygens that combine with each carbon to make CO_2 have a total atomic mass of 32. Thus, the mass of the CO_2 (44) is almost four times the mass of the carbon (12). Thus, your car will emit $4 \times 2000 \text{ kg} = 8 \times 10^3 \text{ kg}$ of carbon dioxide every year.

However, there is more than one car in the US. As we discussed in question 5.1, there are 3×10^8 Americans and about 0.5 car per person. Thus, all American cars emit a total of

$$M_{CO_2} = 8 \times 10^3 \text{ kg/car-year} \times 1.5 \times 10^8 \text{cars}$$
$$= 1 \times 10^{12} \text{ kg/yr}$$

That is 1 billion (10^9) tons.

Now let's compare to reality. According to the US Department of Energy [26], total CO_2 emissions in the US in 2004 were 6 billion tons. Transportation, which includes cars, trucks, railroads, and airplanes, accounted for 2 billion tons of CO_2. That is only twice as much as we estimated for cars alone.

That seems to be a lot of CO_2. Let's again compare it to the mass of the atmosphere. Driving cars will add

$$f = \frac{1 \times 10^{12}\,\text{kg}}{4 \times 10^{18}\,\text{kg}}$$
$$= 2 \times 10^{-7}$$

or 200 parts per billion CO_2 to the atmosphere every year. If you remember those golf balls circling the equator in the first chapter, that would amount to 200 CO_2 golf balls out of the billion needed to circle the globe. This is 1000 times less than natural level of CO_2 in the atmosphere of about 200 parts per million (200 ppm or 2×10^{-4}).

If we want to reduce our share of transportation greenhouse gas emissions, we can either drive less or drive a car with better gas mileage. Driving less means either car pooling* or living closer to work (which restricts our house and neighborhood choices). Driving a more efficient car means either paying more for a hybrid (but not all hybrids get good gas mileage) or driving a smaller car. If you drive an electric car, the electricity still has to be generated somehow, generally by burning fossil fuels. Alternatively, we can pass laws that make *other* people pay the costs of reducing CO_2 emissions. The choice is ours.

* Which carries the risk of developing car pool tunnel syndrome.

Turning gas into trees

10.7

How much carbon dioxide (in kg) does a 1-km^2 new[*] forest absorb per year? Trees absorb carbon dioxide, emit oxygen (thank you, trees!), and use the carbon to make more tree. Assume that the forest is growing in a temperate zone that gets plenty of rainfall (e.g., New Jersey[†] or Germany).

[*] So we do not need to account for trees dying and rotting.
[†] New Jersey may be the "landfill of opportunity," but we assure you that it has some lovely forests too.

HINT: If you cut down all the trees after 20 years and compacted them, how thick a layer would they make on the forest floor?

HINT: Estimate the total mass of all the trees after 20 years.

HINT: The mass of the trees is mostly carbon.

HINT: Average over 20 years.

ANSWER: This is a great problem, because there are four very different ways to solve it. We can estimate (1) the number and size of 20-year-old trees, (2) how thick a layer they would make if we cut them all down and compacted them, (3) the solar energy used by the forest and how much CO_2 would be converted to carbon with that energy, or (4) the water used by the trees and how much cellulose would be made from the hydrogen in the water. Methods (1) and (2) will work much better if we average over 20 years.

The easiest method is to estimate how thick a layer the trees would make if after 20 years we cut the forest down and compacted it. A 20-year-old tree is less than a meter thick. Thus, the layer of biomass will certainly be more than 1 cm (10^{-2} m or 1/2 in.) and less than 1 m, so we will take the geometric mean and use a thickness of 0.1 m (4 in.). Thus, the volume of biomass will be $V = (10^3 \text{ m})^2 \times 0.1 \text{ m} = 10^5 \text{ m}^3$. Since wood has about the same density as water (10^3 kg/m^3), this gives a total mass of $M = 10^3 \text{ kg/m}^3 \times 10^5 \text{ m}^3 = 10^8 \text{ kg}$. Since almost of all of this mass is carbon and since CO_2 has about four times the mass of C, we estimate that 1 km^2 of new forest removes 4×10^8 kg of CO_2 in 20 years or 2×10^7 kg/year of carbon dioxide.

Now let's try the solar energy method. The solar power available at Earth orbit is 10^3 W/m^2. Between night and clouds and the atmosphere, about 10% of that energy is available to the trees. After a billion years of evolution, trees must be more than 1% efficient at using solar energy (but definitely less than 100%) so we'll assume 10% efficiency.* Some of that energy probably goes to maintaining the tree's metabolism so we'll estimate that 50% of the available energy is used to make more tree. Thus, after accounting for night

---

* This is the same efficiency as our best photovoltaic solar cells today.

and clouds (10%), energy conversion efficiency (10%), and energy used for making more tree (50%), only 5×10^{-3} of the 10^3 W/m^2 or 5 W/m^2 is used to make more tree. This gives an available energy

$$E = 5 W/m^2 \times \pi \times 10^7 \text{ s/yr} \times 10^6 \text{ m}^2/\text{km}^2$$
$$\times 6 \times 10^{18} \text{ eV/J} = 10^{33} \text{ eV/yr-km}^2$$

We converted this from joules to electron-volts because extracting the carbon from CO_2 requires 1.5 eV (our standard chemical reaction energy). Thus, we can extract the carbon from 6×10^{32} CO_2 molecules per year. The mass of all this CO_2 is

$$M_{CO_2} = \frac{6 \times 10^{32} \text{ molecules/yr}}{6 \times 10^{23} \text{ molecules/mole}} \times 44 \text{ g/mole } CO_2$$
$$= 4 \times 10^{10} \text{ g/yr} = 4 \times 10^7 \text{ kg/yr}$$

This is a factor of about two more than our previous estimate. Not too bad for such different methods.

Now let's compare to reality. According to the *Encyclopedia of Energy* [27], the average insolation ranges from 100 to 200 W/m^2 and the solar energy capture efficiency ranges from 0.2 to 5%. The highest production levels are from a tree plantation in the Congo, which produces 36 tons/hectare/year or 4×10^6 kg/km^2/yr. Our estimates are about a factor of ten higher even than this maximal production.

Turning trees into gas

10.8

Humans have cleared a lot of forest for agriculture in the last 8000 years. Assuming that all the cleared forest has been burned or otherwise oxidized, how many tons of CO_2 has deforestation added to the atmosphere? How much is this in ppm (parts per million)?

* * * * * * * * * * * * * * *

ANSWER: We need to estimate the area of forested land cleared and the amount of biomass per area. Let's start with the area. Land is 25% of the Earth's surface. We'll estimate that half of all land area was originally forested and half of that has been cleared. Thus, humans have cleared forest from 1/4 of the land area of Earth. This estimate is certainly within a factor of four of reality since we have certainly cleared less than 100% of the land area and more than 1/16 or 6%. The surface area of the Earth (see question 10.1) is

$$A_{Earth} = 4\pi R^2 = 4\pi (6 \times 10^3 \, km)^2 = 4 \times 10^8 \, km^2$$

The land area is one-quarter of that or $10^8 \, km^2$. Humans have cleared the forests from one-quarter of that or $2 \times 10^7 \, m^2$.

Now we need the amount of biomass. We know from the previous question that new forests produce less than $4 \times 10^6 \, kg/km^2/yr$ of biomass. However, after a forest becomes mature, it becomes a steady-state system where the amount of wood that rots (i.e., oxidizes) equals the amount of new wood produced. It takes at least 10 years and less than 100 years for a forest to become mature.* We'll take the geometric mean and use 30 years. Thus, clearing forest for agriculture releases $30 \times 4 \times 10^6 \, kg/km^2 = 10^8 \, kg/km^2$ of biomass. Since CO_2 has four times the mass of carbon, this is $4 \times 10^8 \, kg/km^2$ of CO_2. Before we continue, let's compare this to reality. Typical forest biomass is about $10^7 \, kg$ carbon/km^2 [28]. Thus, our estimate of 10^8 is a factor of 10 too high.

Using the real biomass density, clearing forests released

$$M_{CO_2} = 2 \times 10^7 \, km^2 \times 4 \times 10^7 \, kg/km^2$$

$$= 8 \times 10^{14} \, kg$$

* This definition of maturity refers to the carbon cycle, not to the species of trees in the forest.

of carbon dioxide. That is 10^4 times more than American cars emit each year.

To put it another way, the total mass of the atmosphere is only 4×10^{18} kg. Thus, this is about 200 parts per million (ppm) of the atmosphere. The current atmospheric concentration of CO_2 is about 350 ppm, so that this is a significant input.

Wow! Maybe clearing forests over the last ten thousand years has averted another ice age [29]. Or maybe not.

Risk

Life is dangerous. But some things are far more dangerous than others. Do we really need to worry about shark attacks? Should we be required to put infants in car seats on airplanes? After figuring out these questions, you should be able to see through the scare tactics used against us by politicians and other salesmen.

Chapter 11

Gambling on the road

What is the risk (in the US) of dying per mile traveled in an automobile? What fraction of American deaths are caused by automobiles?

HINT: How many total miles do Americans drive per year? See question 5.1.

HINT: About 4×10^4 Americans are killed on the roads each year.

HINT: The population of the US is about 3×10^8.

HINT: What is our life expectancy?

HINT: If our life expectancy is 100 years, then 1/100 of us will die each year.

ANSWER: We need to estimate the total number of miles traveled and the total number of deaths on the road. In question 5.1 we estimated the total distance that Americans drive each year to be about 2×10^{12} mi. About 4×10^4 Americans die each year in car crashes. Thus, the risk of death is

$$R = \frac{4 \times 10^4 \text{ deaths/yr}}{2 \times 10^{12} \text{ mi/yr}} = 2 \times 10^{-8} \text{ deaths/mi}$$

Wow, there are only two deaths per 100 million miles. That sounds pretty safe.

Let's look at this number a different way. Let's figure out the probability that you, the reader, will die in a car crash. The probability that any one of us will be killed in a car crash is the same as the fraction of Americans who die in car crashes. That is, if one in fifty Americans die in car crashes, then your chance of dying in a car crash is also one in fifty.

Now we need to estimate the number of Americans that die each year. There are 3×10^8 of us. The average American life span is about 75 years (although we plan to live much longer than that). Thus, one in every 75 Americans dies every year. This means that the number of deaths every year is

$$N_d = \frac{3 \times 10^8 \text{ Americans}}{75 \text{ yr}} = 4 \times 10^6 \text{ deaths/yr}$$

Thus, the total lifetime probability of dying in a car crash is

$$P_{crash} = \frac{4 \times 10^4 \text{ car deaths/yr}}{4 \times 10^6 \text{ deaths/yr}} = 0.01$$

or 1%. Thus, 1% of us will die in a car crash. (This is the total probability starting at age zero. Since you, gentle reader, are probably slightly older than that, your probability of dying in a car crash is somewhat reduced. Whew!)

The plane truth

What is the risk in America of dying per mile traveled in a large airplane? How does this compare to the risks of driving?

HINT: How many total miles do Americans fly each year?

HINT: How many airplane flights do we take? See the introduction.

HINT: How far is each flight?

HINT: How often do large planes crash?

HINT: How many people die per crash?

ANSWER: We need to estimate the total miles traveled and the total deaths per year. We estimated in the introduction that Americans take 7×10^8 flights per year. The average distance per flight is probably more than 300 miles (it is almost easier to drive) and less than 3000 miles (the distance from NY to LA), so we will use the geometric mean of 1000 miles. Thus, Americans fly a total distance of $d = 7 \times 10^8$ flights $\times 10^3$ miles/flight $= 7 \times 10^{11}$ mi. This is 1/3 as far as we drive.

Large planes do not crash every year. Whew! The crash frequency is less than once a year and more than once a decade. We'll take the geometric mean of once every three years (since $3 = \sqrt{1 \times 10}$). When a plane crashes, it kills about 100 people. Thus, large plane crashes kill about 30 people per year. This means that the probability of dying in a large plane crash per mile is

$$R = \frac{30 \text{ deaths/yr}}{7 \times 10^{11} \text{ mi/yr}} = 4 \times 10^{-11} \text{ deaths/mi}$$

That is almost 1000 times safer per mile than driving.

Note that expressing airplane fatalities in terms of deaths per mile is somewhat misleading since almost all crashes occur during takeoff and landing. A 3000-mile plane flight is almost exactly as dangerous as a 300-mile plane flight.

This means that some airline safety regulations could be misguided. For example, if we required infants to ride in infant seats on airplanes, it could cost lives. This would make plane travel more expensive, since parents would have to pay for a plane ticket for their infant. Thus, fewer families would fly and more would drive. Because driving is so much more dangerous than flying, many more people would be killed on the roads than would be saved by using infant seats on the airplane.

HINT: How many people are killed by sharks each year in the US?

HINT: Shark attacks are so newsworthy, you can assume that every death is reported nationwide.

HINT: How many people are at the beach on a typical summer day?

HINT: How many typical summer days are there?

HINT: How far do people drive to the beach?

HINT: Driving kills 2×10^{-8} people per mile.

* * * * * * * * * * * * * * * *

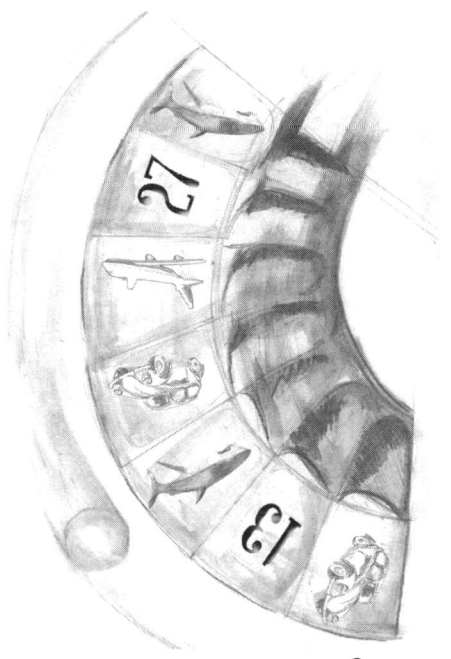

Compare the risks of getting killed by a shark at the beach and of driving to the beach.

Life's a beach

11.3

ANSWER: Shark attacks are scary. Shark attacks are newsworthy. Just about every person killed by a shark will make the national news. We read about shark deaths at the beach every few years. This indicates that there is about one fatality per year. (To humans, that is. We kill lots more sharks per year than that.) In reality, the average for the US is about 0.5 per year.

Now we need to estimate the number of beach goers per year. About 10% of the US population lives within 20 miles of the coast (it's certainly between 1 and 100%). About 10% of them go to the beach on any given summer day. This means that in July and August, $3 \times 10^8 \times 0.1 \times 0.1 = 3 \times 10^6$ Americans are at the beach on any given day. Since there are 60 days in July and August, this means that there are $60 \times 3 \times 10^6 = 2 \times 10^8$ beach visits each year. This is a bit low but reasonably close to reality since California alone has 10^8 beach visits per year [30].

Each beach goer probably drives about 10 miles each way to get to the beach (more than one and less than 100). The total distance driven by beach goers is thus about $d = 20\,\text{mi} \times 2 \times 10^8$ visits $= 4 \times 10^9$ mi. This will cause an average of

$$N_{\text{deaths}} = 4 \times 10^9\,\text{mi} \times 2 \times 10^{-8}\,\text{deaths/mi} = 80\,\text{deaths}$$

Thus, driving to the beach is about 100 times more dangerous than shark attacks. Despite this, or perhaps because of this, the shark attacks get far more media coverage and publicity.*

---

* Driving accidents are just so, you know, pedestrian.

Up in smoke

11.4

On average, how much does each cigarette smoked by a heavy smoker shorten his or her life expectancy?

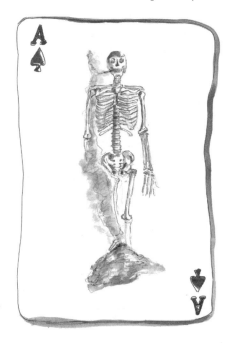

HINT: How many cigarettes do heavy smokers smoke?

HINT: How many years of life does the average smoker lose?

ANSWER: We need to estimate how many cigarettes a heavy smoker smokes and how many years of life he or she will lose. We will do this for the average smoker. Your Uncle Joe may have smoked like a chimney until he was shot by a jealous lover at age 97 while skydiving, but he got lucky. Someone will get lucky and win the lottery, but the average lottery gambler loses money. Some people live long lives despite smoking, but the average smoker dies early.

Smoking kills primarily through lung cancer and heart disease. These are late-onset diseases. They typically start killing at age 50 or so. The average smoker must lose much more than one year of life* and less than 30 years (since smoking starts to kill at about age 50 and life expectancy is less than 80). Taking the geometric mean of 1 and 30, smokers die five years earlier than nonsmokers.

If you started smoking at age 18 (when it became legal to buy cigarettes) and continued until you died at age 70, at one pack of cigarettes per day, you would have smoked

$$N = 50 \text{ years} \times 400 \text{ days/year} \times 20 \text{ cigarettes/day}$$

$$= 4 \times 10^5 \text{ cigarettes}$$

If we make the totally ridiculous assumption that each cigarette makes the same contribution to mortality, then each cigarette will cost you

$$t = \frac{5 \text{ years}}{4 \times 10^5 \text{ cigarettes}} = 10^{-5} \text{ years}$$

$$= 10^{-5} \text{ years} \times \pi \times 10^7 \text{ s/yr} = 300 \text{ s} = 5 \text{ min}$$

Thus, each cigarette would cost you about the amount of time it takes to smoke it.

---

* If a smoker only lost one year of life, we would not make so much of a fuss about smoking.

Comparing to reality,* a study in the *British Medical Journal* [31] found that there is a difference in life expectancy between smokers and nonsmokers of 6.5 years and the average smoker consumes a bit less than a pack per day. They found that each cigarette costs an average of 11 min of life. Maybe we should have published.

Note that the assumption that each cigarette does an equal amount of damage is both unverifiable and probably nonsense. Biological systems frequently have thresholds below which no damage is done. One ton of rocks will crush you. However, the first pound or kilogram of rock will not have any adverse affect at all. Similarly, swallowing two tablets of acetaminophen will help your headache but swallowing the entire bottleful will kill you.

---

* Or at least to a published study.

Unanswered Questions

There are many, many more questions that could be asked and answered. Just look at the world around you and apply your newfound skills. Here is a set of unanswered questions to get you started.

Chapter 12

* * * * * * * * * * * * * * *

1. How many 1-gal (4-L) buckets of water are needed to empty Loch Ness (or Lake Erie)?

2. How many cigarettes are smoked annually in the US? If you place them end-to-end, how far will they stretch?

3. How many video rental stores are there in the US (or Europe)?

4. How many people are talking on their cell phones at this instant?

5. How many people are eating lunch at this instant?

6. How fast does human hair grow (in m/s or mph)?

7. How many grains of sand are there in all the beaches of the world?

8. How many blades of grass are there on the average (natural) football field?

9. If all the lottery tickets sold in the US in one year were stacked up, how tall would the stack be?

10. How much total time do Americans spend driving in one year? Express your answer in hours, years, and lifetimes.

11. What is your average bicycle travel speed? Include the time you spend riding your bicycle and the time you spend earning money to pay for your bicycle.

12. How much gasoline would the US save if the speed limit was lowered from 65 to 55 mph? How many extra hours of the time would be spent driving for each gallon of gas saved?

13. How much rubber (in kg) is deposited on American roadways every year by automobile tires?

14. What is the potential energy of a large raindrop in a cloud?

15. You crash your car into highway crash barrels (large barrels filled with sand) at highway speed. What force is exerted on your body as you stop? (Assume that you are wearing your seat belt and your air bag deploys properly.)

16. How much potential energy does the landfill referred to in chapter 3 have after Americans have dumped all their garbage there for 100 years?

17. What is the kinetic energy of a rifle bullet?

18. What force does a catcher exert on a baseball when he catches a fast pitch?

19. How much energy can be released by eating 1 kg of chocolate chip cookies?

20. What is the total kinetic energy of all the vehicles on the road in the United States (or Europe) at this instant?

21. If you drop a water ballon from a 30-story building, with what speed will it hit the ground?

22. How much does a human sweat in a hot climate ($T = 98.6\,°F = 37\,°C$)? Express your answer in liters/day. (Assume that all of the energy you ingest can be removed only by sweating. Water absorbs about 1000 J/g when it evaporates.)

23. How much food (in kg) does a typical human consume in her lifetime? How does that compare to her mass?

24. What is the kinetic energy of the Moon as it orbits the Earth?

25. What is the energy density of ^{235}U (in J/kg)?

26. If every American switched to driving an electric car we would use more electrical energy.

How many 1-GW power plants would it take to supply that much electrical energy?

27. How many tons of hydrogen are converted to helium in the Sun every year?

28. What proportion of the Sun's mass is converted to energy during the 10-billion year lifetime of the Sun? (Assume that its power output is relatively constant over its lifetime.)

29. How many miles of roads does the US (or Europe) have?

30. How much carbon dioxide (in kg) does one person emit into the atmosphere each year? All humans?

31. How much force does the atmosphere exert on the front of your body?

32. How large a helium ballon (in m^3) would be needed to lift your car? Helium has about one-tenth the density of air.

33. How much, on average, will each high-fat meal consumed shorten your life?

Appendix A
Needed Numbers and Formulas

Although it is extraordinary how much can be deduced from just a little knowledge, a little knowledge is necessary. We do not believe that it is dangerous. Here are the important numbers and formulas we use in this book.

A.1 Useful Numbers

US population (2006) 3×10^8

World population (2006) 6×10^9

Number of items/mole (Avogadro's number) 6×10^{23}

Electron volts/joule (eV/J) 6×10^{18}

1 m/s   2 mph

1 year $\pi \times 10^7$ s

Size of an atom 10^{-10} m

Radius of the Earth 6×10^6 m

Earth–Sun distance 1.5×10^{11} m

1 Calorie 4×10^3 J

Chemical reaction energy   1.5 eV

Mass of one mole of carbon   12 g

Acceleration of gravity at Earth's surface g $10\,\text{m/s}^2$

A.2 Handy formulas

Potential energy, PE mgh

Mass   volume × density

Kinetic energy, KE $\frac{1}{2}mv^2$

Work   force × distance

Work   change in KE

Energy   power(W) × time(s)

A.3 Metric Prefixes

Size	Prefix	Abbreviation
10^9	giga	G
10^6	mega	M
10^3	kilo	k
10^{-2}	centi	c
10^{-3}	milli	m
10^{-6}	micro	μ
10^{-9}	nano	n

Appendix B
Pegs to Hang Things On

Length in meters (m)	Object
10^{11}	Earth–Sun distance (1.5×10^{11} m)
10^7 (10^4 km)	Earth's diameter (8000 mi, or 1.3×10^4 km)
10^6 (10^3 km)	Distance from New Orleans to Detroit (1600 km)
10^5 (10^2 km)	Lake Michigan (length)
10^4 (10 km)	Mt. Everest (height)
10^3 (1 km; 0.6 mi)	George Washington Bridge
10^2	Football field (length)
10^1	Tennis court
10^0	Tall man's stride
10^{-1} (10 cm)	Person's hand (width)
10^{-2} (1 cm)	Sugar cube
10^{-3} (1 mm)	Coin (thickness)
10^{-4}	Human hair (thickness)
10^{-5}	Human cell (diameter)
10^{-6} (1 micron [1 μm])	Soap-bubble film (thickness)
10^{-9} (1 nanometer [1 nm])	Small molecule
10^{-10}	Atom

Area in square meters (m^2)	Typical object
10^{14}	Land area of the Earth
10^{12}	Egypt; Texas
10^{11}	New York State; Iceland
10^{9}	Los Angeles; Virginia Beach
10^{8}	Manhattan
10^{6} (1 km^2)	City of London
10^{4}	Football field
10^{2}	Volleyball court
10^{0}	Small office desk
10^{-4} (1 cm^2)	Sugar cube (one side only)
10^{-6} (1 mm^2)	Head of a pin
10^{-8}	Pixel on computer display

Density in kilograms per cubic meter (kg/m^3)	Item
10^{18}	Neutron star; atomic nucleus
10^{9}	White dwarf star
10^{4}	Lead; iron
10^{3} (1 ton/m^3, 1 kg/L, 1 g/cm^3)	Water; human body
10^{0}	Earth's atmosphere at sea level

Mass in kilograms (kg)	Object
10^{30}	The Sun
10^{27}	Jupiter
10^{25}	Earth
10^{21}	Earth's oceans
10^{18}	Earth's atmosphere
10^{15}	World coal reserves (estimated)
10^{12}	World oil production in 2001
10^{11}	Total mass of human world population
10^{10}	Great Pyramid of Giza
10^{9}	Matter converted into energy by the Sun each second
10^{8}	Aircraft carrier
10^{7}	RMS *Titanic*
10^{6}	Launch mass of the space shuttle
10^{5}	Largest animal, the blue whale
10^{4}	Large elephant
10^{3} (1 ton)	Automobile (small)
10^{2}	Lion; large human
10^{1}	Microwave oven; large cat
10^{0}	1 liter or quart of water
10^{-1}	Human kidney; apple; rat
10^{-2}	Lethal dose of caffeine; adult mouse; large coin
10^{-3} (1 g)	Sugar cube
10^{-4}	Caffeine in a cup of coffee
10^{-6} (1 mg)	Mosquito
10^{-7}	Lethal dose of ricin
10^{-9} (1 μg)	Sand grain (medium)
10^{-12} (1 ng)	Human cell
10^{-27}	Neutron; proton; hydrogen atom
10^{-30}	Electron

Bibliography

[1] A. Asaravala. Landing a job can be puzzling. *Wired.com*, June 2003. http://www.wired.com/news/culture/0,1284,59366,00.html.

[2] J. Kador. *How to Ace the Brain Teaser Interview*. McGraw-Hill, New York City, 2004. http://www.jkador.com/brainteaser/

[3] W. Poundstone. *How Would You Move Mount Fuji?* Little, Brown, New York, 2003.

[4] C. Swartz. First, the answer. *The Physics Teacher* 33:488, 1995.

[5] P. Morrison. Fermi questions. *American Journal of Physics*, 31:626, 1963.

[6] T. Isenhour. Private communication. Originally attributed to H. Jeffries.

[7] G. Lucier and B.-H. Lin. Americans relish cucumbers. *Agricultural Outlook*, page 9, 2000. www.ers.usda.gov/publications/agoutlook/dec2000/ao277d.pdf

[8] T. L. Isenhour and L. G. Pedersen. *Passing Freshman Chemistry*. Harcourt Brace Jovanovich, New York, 1981.

[9] Basic facts: municipal solid waste (US). Technical report, US Environmental Protection Agency, Washington, DC, 2006. http://www.epa.gov/msw/facts.htm

[10] P. Jillette and Teller. Recycling. *Bullsh*t!* 2004.

[11] Stanford Comprehensive Cancer Center. Stanford bmt body surface area calculator, 2006. http://bmt.stanford.edu/calculators/bsa.html

[12] *McGraw-Hill Encyclopedia of Science & Technology*. McGraw-Hill Professional, New York, 2004.

[13] G. Elert. The physics hypertextbook, 2006. http://hypertextbook.com/facts/1998/StevenChen.shtml

[14] N. Juster. *The Phantom Tollbooth*. Knopf, New York, 1961.

[15] G. Elert. The physics hypertextbook, 2006. http://hypertextbook.com/physics/matter/energy-chemical/

[16] I. Buchmann. *Batteries in a Portable World*. Cadex Electronics, Richmond, BC, Canada, 2001. http://www.batteryuniversity.com/index.htm

[17] L. Brown. *Plan B: Rescuing a Planet Under Stress and a Civilization in Trouble*. W.W. Norton, New York, 2003. http://www.earth-policy.org/Books/PB/ PBch8_ss4.htm

[18] M. S. Matthews T. Gehrels, and A. M. Schumann, editors. *Hazards Due to Comets and Asteroids*. University of Arizona Press, Tucson, AZ, 1994. http://seds.lpl.arizona.edu/nineplanets/nineplanets/ meteorites.html

[19] International energy outlook 2006. Technical Report DOE/EIA-0484(2006), DOE Energy Information Administration, Washington, DC, 2006. http://www.eia.doe.gov/oiaf/ieo/world.html

[20] W. Thomson. On the age of the sun's heat. *Macmillan's Magazine* 5:288, 1862. Reprinted in *Popular Lectures and Addresses*, vol. 1, by Sir William Thomson. Macmillan, London, 1891.

[21] Sea level and climate. Technical Report Fact Sheet 002-00, US Geological Survey, Washington, DC, 2000. http://pubs.usgs.gov/fs/fs2-00/

[22] *The World Factbook 2006*. Central Intelligence Agency, Office of Public Affairs, Washington, DC, 2006. https://www.cia.gov/cia/publications/factbook/ index.html

[23] Danish Wind Industry Association. Wind turbine power calculator. http://www.windpower.org/en/tour/ wres/pow/index.htm

[24] C. D. Elvidge et al. U.S. constructed area approaches the size of Ohio. *Eos, Trans. AGU* 85:233, 2004. http://www.agu.org/pubs/crossref/2004/2004EO240001.shtml

[25] G. Grimvall. Socrates, Fermi, and the second law of thermodynamics. *American Journal of Physics*, 72:1145, 2004.

[26] Emissions of greenhouse gases in the United States 2004. Technical Report DOE/EIA-0573(2004), DOE Energy Information Administration, Washington, DC, 2005. http://www.eia.doe.gov/oiaf/1605/ggrpt/carbon.html

[27] D. L. Klass. Biomass for renewable energy and fuels. In *Encyclopedia of Energy*. Elsevier, Amsterdam, 2004. www.bera1.org/cyclopediaofEnergy.pdf

[28] R. A. Houghton. Aboveground forest biomass and the global carbon balance. *Global Change Biology* 11:945, 2005. www.whrc.org/resources/published_literature/pdf/HoughtonGCB.05.pdf

[29] W. Ruddiman. *Plows, Plagues and Petroleum*. Princeton University Press, Princeton, NJ, 2005.

[30] G. H. Burgess. California beach attendance, surf rescues, and shark attacks 1990–2000. International Shark Attack File, 2005. http://www.flmnh.ufl.edu/fish/Sharks/statistics/CAbeachattacks.htm

[31] M. Shaw, R. Mitchell, and D. Dorling. Time for a smoke? One cigarette reduces your life by 11 minutes. *British Medical Journal* 320:53, 2000. http://bmj.bmjjournals.com/cgi/content/full/320/7226/53

Index